Statistical Framework for Recreational Water Quality Criteria and Monitoring

STATISTICS IN PRACTICE

Advisory Editors

Stephen Senn
University of Glasgow, UK
Marian Scott
University of Glasgow, UK

Founding Editor

Vic Barnett
Nottingham Trent University, UK

Statistics in Practice is an important international series of texts which provide detailed coverage of statistical concepts, methods and worked case studies in specific fields of investigation and study.

With sound motivation and many worked practical examples, the books show in down-to-earth terms how to select and use an appropriate range of statistical techniques in a particular practical field within each title's special topic area.

The books provide statistical support for professionals and research workers across a range of employment fields and research environments. Subject areas covered include medicine and pharmaceutics; industry, finance and commerce; public services; the earth and environmental sciences, and so on.

The books also provide support to students studying statistical courses applied to the above areas. The demand for graduates to be equipped for the work environment has led to such courses becoming increasingly prevalent at universities and colleges.

It is our aim to present judiciously chosen and well-written workbooks to meet everyday practical needs. Feedback of views from readers will be most valuable to monitor the success of this aim.

A complete list of titles in this series appears at the end of the volume.

Statistical Framework for Recreational Water Quality Criteria and Monitoring

Edited by

Larry J. Wymer

US Environmental Protection Agency

John Wiley & Sons, Ltd

Other Wiley Editorial Offices

John Wiley & Sons Inc., 111 River Street, Hoboken, NJ 07030, USA

Jossey-Bass, 989 Market Street, San Francisco, CA 94103-1741, USA

Wiley-VCH Verlag GmbH, Boschstr. 12, D-69469 Weinheim, Germany

John Wiley & Sons Australia Ltd, 42 McDougall Street, Milton, Queensland 4064, Australia

John Wiley & Sons (Asia) Pte Ltd, 2 Clementi Loop #02-01, Jin Xing Distripark, Singapore 129809

John Wiley & Sons Canada Ltd, 6045 Freemont Blvd, Mississauga, ONT, L5R 4J3

Wiley also publishes its books in a variety of electronic formats. Some content that appears in print
may not be available in electronic books.

Anniversary Logo Design: Richard J. Pacifico

British Library Cataloguing in Publication Data

A catalogue record for this book is available from the British Library

ISBN-13 978-0470-03372-2 (HB)

Typeset in 10/12pt Times by Integra Software Services Pvt. Ltd, Pondicherry, India
Printed and bound in Great Britain by TJ International, Padstow, Cornwall
This book is printed on acid-free paper responsibly manufactured from sustainable forestry in which at
least two trees are planted for each one used for paper production.

Contents

Contributors

ALEXANDRIA B. BOEHM Department of Civil and Environmental Engineering, Stanford University, Stanford, CA, USA.

G.F. CRAUN Gunther F. Craun & Associates, Staunton, VA, USA.

ALFRED DUFOUR National Exposure Research Laboratory, US Environmental Protection Agency, Cincinnati, OH, USA.

A.H. EL-SHAARAWI National Water Research Institute, Environment Canada, Burlington, Ontario, Canada.

S.R. ESTERBY Mathematics, Statistics and Physics, Irving K. Barber School of Arts and Sciences, University of British Columbia Okanagan, Canada.

ALESSANDRO FASSÒ Department of Information Technology and Mathematical Methods, University of Bergamo, Dalmine BG, Italy.

RICHARD O. GILBERT Environmental Statistician, Rockville, MD, USA.

DEYI HOU Department of Civil and Environmental Engineering, Stanford University, Stanford, CA, USA.

GRAHAM MCBRIDE National Institute of Water and Atmospheric Research, Hamilton, New Zealand.

MEREDITH B. NEVERS Lake Michigan Ecological Research Station, United States Geological Survey, Porter, IN, USA.

D.F. PARKHURST Environmental Science Research Center, School of Public and Environmental Affairs, Indiana University, Bloomington, IN, USA.

STEPHEN SCHAUB Office of Water, US Environmental Protection Agency, Washington, DC, USA.

J.A. SOLLER Soller Environmental, Berkeley, CA, USA.

TIMOTHY J. WADE National Health and Environmental Effects Research Laboratory, US Environmental Protection Agency, Research Triangle Park, NC, USA.

STEPHEN B. WEISBERG Southern California Coastal Water Research Project, Costa Mesa, CA, USA.

RICHARD L. WHITMAN   Lake Michigan Ecological Research Station, United States Geological Survey, Porter, IN, USA.

ALBRECHT WIEDENMANN   Infectious Disease Control and Environmental Hygiene, Public Health Office, Administrative District Esslingen, Germany.

LARRY J. WYMER   National Exposure Research Laboratory, US Environmental Protection Agency, Cincinnati, OH, USA.

Preface

Since its creation in 1970, the United States Environmental Protection Agency (EPA) has been concerned with the microbial quality of the nation's recreational waters, one of the responsibilities which it inherited from its predecessors, in particular, the Federal Water Quality Administration of the Department of the Interior. In 1972, Congress passed amendments to the Water Pollution Control Act of 1948, consolidating the authority for its implementation to the Administrator of the EPA. These amendments collectively have become known as the Clean Water Act and directed the EPA to consider the costs of discharges with respect to impairment of their various uses, including recreation.

More recently, another round of amendments to the Act, the Beaches Environmental Assessment and Coastal Health (BEACH) Act of 2000, was passed by Congress and reflects continued importance of recreational water quality. In the BEACH Act we see the emergence of a new concern, namely *timeliness* in the determination of water quality. The development of molecular methods of detecting and quantifying microbes, such as rapid polymerase chain reaction (PCR) techniques, has opened the possibility of nearly 'real-time' monitoring of beaches, where previously one had to wait a sufficient length of time, as long as 24–48 hours, for bacteria to grow in broth or on culture medium.

A combination of innovative microbiological methods, advances in statistical theory, and increased concern for the accurate characterization and forecasting of recreational water quality has created the need for ever more sophisticated statistical treatment of recreational monitoring. The availability of same-day results from microbial monitoring means that we are no longer tied to a control chart type of approach to monitoring as in the past. The option of accurately characterizing 'today's' beach water quality is now a possibility, and with that comes the statistical complexity of the relationships between sampling techniques, sample size, and how to utilize sample data for decision making.

Conversations between the EPA Office of Research and Development (ORD), of which I am a part, and the EPA Office of Water (OW), which is charged with setting criteria in accordance with the BEACH Act of 2000, have made it clear that statistical guidance for such criteria is needed. In December 2004, a workshop was conducted at ORD's facilities in Cincinnati, Ohio, which attempted to address the many statistical concerns expressed by OW, and it is this workshop that provided the idea to publish a book on the subject. All of the participants in that workshop have contributed chapters, in addition to several others who were asked, and graciously accepted, to contribute chapters based on the need for guidance in their respective areas of expertise.

We begin with an account of the development of water quality standards in the United States during the twentieth century. Chapter 1 covers the fascinating history of quantitative microbiology and epidemiology related to water quality measurement via indicator bacteria, starting with an early application of maximum likelihood to a random dispersion model for fermentation tube experiments by McCrady in 1915.

Chapter 2 discusses recreational water monitoring in the context of management decision making. Monitoring can be used for making decisions with regard to whether health warnings should be issued, whether compliance goals for treated wastewater discharges are being met, and whether previous actions taken to improve or protect water quality are effective, and these different uses all entail different demands on sampling design and data quality objectives.

Chapter 3 delves into the issue of indicator organisms in recreational water monitoring, the basis for their use, what constitutes a good indicator, and their relationship to what really matters at a bathing beach – human health. Relationships between pathogens and indicators are examined and alternative indicators, other than the commonly used fecal coliforms, *E. coli*, and enterococci (fecal streptococci) discussed.

The discussion in Chapter 4 turns to alternate philosophies for beach sampling: process control and acceptance sampling. In the case of the former, we question whether the beach is behaving over some long term as one would expect of a normal, safe beach, while in the latter we question whether the 'product' (i.e. the bathing beach water on that day) meets our specifications. As mentioned earlier, it is the development of molecular methods for the rapid assessment and quantification of indicator bacteria that have enabled us to consider an acceptance sampling approach.

Various sampling strategies, including simple random, stratified, systematic, composite, and ranked set sampling, are presented in Chapter 5. Model-based sampling is discussed as an alternative to random selection of samples. Application of the design-based and model-based approaches to microbiological estimation are illustrated and discussed.

In response to concerns that have been raised about using geometric means to characterize microbiological populations in cases where arithmetic means may be more appropriate, Chapter 6 addresses this problem specifically for the case of beach monitoring for the protection of human health. We examine the relative merits of the two types of means, and in the process discuss some properties of geometric means, 'log means' (the mean of a series of logarithms of numbers), arithmetic means, and lognormal distributions.

In Chapter 7 we present a case study of an intensive sampling research project conducted by the EPA in 2000 among five beaches. These beaches were selected for their diverse ecologies, representing East Coast and West Coast marine beaches, an estuarine beach, a riverine beach and a freshwater beach. We present a reasonable probability mass function to describe the distribution of indicator bacteria among samples taken at the beach and to determine systematic differences among the fixed sampling locations at each beach. Given that a lognormal approximation is often used to describe such data, we examine the adequacy of this approximation and derive the appropriate estimates for the corresponding 'log variances'.

Chapters 8 and 9 are primarily concerned with risk assessment as related to microbial quality of recreational water. Chapter 8 discusses epidemiological studies and distinctions among various such studies that have been conducted around the world, and the relatively new area of quantitative microbial risk assessment, which has the goal of characterizing risk based on exposure to specific pathogens. A specific design for epidemiological studies, randomized controlled trials, that is particularly popular in the European community, is detailed in Chapter 9.

Microbial water quality modeling is covered in Chapter 10. Here, predictive models based on such variables as water conditions and meteorological data are presented as alternatives or adjuncts to sampling to obtain more timely information. 'Nowcasting' is proposed to determine today's water quality based on its recent past. Case studies of recreational water quality models for a Great Lakes and a West Coast beach are given.

Finally, Chapter 11 discusses sensitivity analysis and design of experiments as tools for developing and refining computer models for water quality and assessing their output. Sources of uncertainty, such as model inputs and output, their determination and initial setup, and ways to assess their relative influence are described.

Each chapter has been subjected to peer review. Specifically for Chapters 1, 6, and 7, which were authored by EPA researchers:

The United States Environmental Protection Agency through its Office of Research and Development funded and managed the research described here. It has been subjected to Agency's administrative review and approved for publication.

The companion website for the book is http://www.wiley.com/go/waterquality.

1

The evolution of water quality criteria in the United States, 1922–2003

Alfred Dufour and Stephen Schaub

National Exposure Research Laboratory, US Environmental Protection Agency, Cincinnati, OH, and Office of Water, US Environmental Protection Agency, Washington, DC

The microbiological quality of recreational waters was first discussed in the United States as early as 1922 by the American Public Health Association's Committee on Bathing Beaches (APHA, 1922). The Committee surveyed 2000 physicians and state health officials inquiring about the prevalence of infections associated with bathing places. The Committee Report in 1924 (APHA, 1924) reviewed the survey results and concluded that there was not enough evidence to develop bathing water standards for natural waters. In June 1933 the Joint Committee on Bathing Places was formed and in their first report noted that because of the great lack of epidemiological information no bacterial standards were adopted (APHA, 1933). They also stated that they did not want to propose arbitrary standards or measures that might promote public hysteria about the dangers of outdoor bathing places. By 1936 they were still not convinced that bathing places were a major health problem and re-stated their position on developing bacterial standards for bathing places.

Statistical Framework for Recreational Water Quality Criteria and Monitoring Edited by Larry J. Wymer
© 2007 John Wiley & Sons, Ltd

The reluctance to propose bacterial standards for outdoor bathing places was again evident in 1936, 1940, and 1955 (APHA, 1936, 1940, 1957). The Committee did attempt to find evidence of the risk of illness from bathing waters prior to their reports in 1940 and 1955, but they found no compelling evidence. They stated that very little reliable data were available to implicate bathing places in the spread of disease (APHA, 1957).

Various states, meanwhile, began to put in place water quality criteria for their bathing places. These criteria ranged from 50 coliforms per 100 ml to 2400 coliforms per 100 ml.[1] The great diversity in state standards for bathing beach waters likely resulted from a lack of guidance from federal authorities or national organizations with interests in this area. In 1963, the Sanitary Engineering Division of the American Society of Civil Engineers, in a progress report (Senn, 1963), included an appendix on the then current state standards for the bacteriological quality of surface waters used for bathing, compiled by William F. Garber. This list is very informative on the various standards and approaches taken to monitor the quality of bathing beach waters. It includes a survey of all 50 states, the standard they used for controlling surface water quality, how the standard was calculated, and whether they used a sanitary survey in conjunction with their standard. Only 12 states had no standard for bathing beach waters in 1963. The other 38 states had standards which differed significantly from each other.

Among those states that did have microbiological standards for water quality, all used coliforms as the measure of that quality and all used the most probable number (MPN) procedure to estimate the number of coliform bacteria in a water sample. The number of coliforms used as a limiting value for water quality ranged from 50 per 100 ml to 2400 per 100 ml, with about half of the states using 1000 coliforms per 100 ml as the limiting value. Over half of the states calculated the upper limit using the arithmetic average of the MPN estimates. Approximately 23 % of the states used either the geometric mean or the median to calculate the threshold value. About 13 % of the states used a single threshold value, such as a percentile value.

Some interesting facts emerge from this listing of state bacteriological water quality standards that were in place in 1963, the most significant of which was the fact that they were all put in place without the benefit of health data showing a relationship between health effects in swimmers and water quality. At that time little information was available from epidemiology studies or from waterborne outbreaks of disease or case studies. Another interesting fact was the broad use of the arithmetic mean as a water quality standard prior to 1963, in contrast to the almost universal use of the geometric mean today. The use of the arithmetic mean at that time may have been due to suggestions by Thomas (1955) and Pomeroy

---

[1] The term 'coliform' was used in the referenced papers to describe bacteria that were believed to be always associated with fecal contamination. The term was changed to 'total coliform' around 1965 (Orland, 1965) when 'fecal coliform' was introduced for a subset of the coliforms The fecal coliforms were thought to be more specific to fecal pollution. *Escherichia coli*, one of a few of the microorganisms that comprise the fecal coliform group, is used today because it is even more specific for identifying the presence of feces in water. Hereafter, in this chapter, the term 'coliforms' shall imply what we currently call 'total coliforms'.

(1955). Thomas indicated that the arithmetic mean of a time series of coliform density determinations is superior to other commonly used averages, such as the geometric mean and the median, as a measure of health hazards associated with enteric pathogens such as *Salmonella* and *Shigella*. Thomas' suggestion was based on the premise that the highest risk of infection was related to the highest values of coliform measured and that the median and geometric mean had a tendency to minimize the high values, resulting in an underestimate of the true risk. Discussions about the use of the geometric mean versus the arithmetic mean continue to the present day with regard to the setting a standards for recreational waters (Parkhurst, 1998). The current use of the geometric mean to set standards for water quality is based mainly on the fact that most environmental water quality measurements, after transformation to logarithms, have a normal distribution and that this normal distribution lends itself to the use parametric statistics for evaluating data.

Most of the states established their water quality standards on an individual basis, while other states used approaches developed by other states. There was no distinct pattern to the development of standards. Three examples of the diversity of setting water quality standards are given by the states of Connecticut, the Ohio River Valley Sanitation Commission, located in Cincinnati, Ohio, and California.

Connecticut surveyed their entire Long Island Sound coastline, taking samples every 1000 ft. Four classes of water quality were developed from the data. Class A included 0–50 coliforms per 100 ml. Class B included coliform densities from 51 to 500 per 100 ml, and Class C ranged from greater than 500 to 1000 coliforms per 100 ml. Class D included coliform densities greater than 1000 per 100 ml. Analysis of the miles of coastline that contain coliform densities less than 1000 per 100 ml indicated that this level could be obtained about 90 % of the time (Scott, 1932). This classification of waters also correlated very well with the sanitary survey conducted during the shoreline water survey. Connecticut used the information to set a water quality standard of less than 1000 coliform per 100 ml. Water samples containing less than 1000 coliforms per 100 ml were graded as acceptable. This approach might be looked on as the attainment approach.

Streeter (1951), an engineer associated with the Ohio River Sanitation Commission, used an analytical approach to develop water quality criteria. He used data reported by Kehr and Butterfield (1943) who showed the relationship between typhoid morbidity and the ratio of coliforms to *E. typhosa* (*S. typhosa*). For morbidity rates in the Ohio River Valley the estimated ratio was about one *S. typhosa* to 170 000 coliforms. In addition to this ratio, Streeter used the number of bathers per day (1000 in his calculations), the estimated volume of water swallowed by a swimmer (10 ml) and the average coliform content of the bathing water per milliliter. His calculations indicated that over a 90-day swimming season, if a bather swam every day, his risk of exposure to one *S. typhosa* would-be approximately 1 in 19. Kerr and Butterfield had estimated that about one in every 50 persons who ingested a single *S. typhosa* bacterium would actually contract typhoid fever, and this was used to estimate that over a 90-day season the risk would be about 1 in 950 exposures for an individual. Streeter considered this to be a very remote hazard. He concluded that water meeting a bathing water standard of 1000 coliforms per 100 ml would not present a great hazard to a single bather or group of bathers.

California's Bureau of Sanitary Engineering (1943) used a somewhat different approach to setting standards for marine bathing waters. A limit of 10 *E. coli* per milliliter was considered to assure safety for recreational use of surf waters. This standard was based on a number of considerations. First, the limit was about 500 times greater than the treated drinking water standard, which was well accepted at that time. Second, there were no epidemiological studies to indicate that waters within this standard were associated with health effects. Third, natural bodies of water seldom exceeded this level. Fourth, the limit seems to be based on common sense and therefore acceptable to the public. It is perhaps coincidental that these three states, even though they used different approaches in the development of bathing water standards, all arrived at the same unacceptable level of bacteriological water quality, namely, 1000 coliforms per 100 ml of sample for both fresh and marine waters.

Much of the reluctance on the part of state and local authorities to set bathing water quality standards was due to the lack of evidence concerning the level of microbial pollution that was detrimental to the health of bathers. In order to remedy this lack of information, the US Public Health Service (PHS), in 1948, initiated a series of studies to obtain data on the relationship between bathing water quality and illness among swimmers (Stevenson, 1953). The first of three studies was conducted at two Lake Michigan beaches in Chicago. The investigators were unable to show an excess illness in swimmers at either of the beaches when the median coliform density was equal to or less than 190 per 100 ml. Further analysis was performed on selected data from both beaches. The three highest and three lowest coliform values were selected from each beach to determine if there was a difference in the gastrointestinal illness incidence rates A statistically significant difference in the illness incidence rate was observed only at one beach where the mean coliform density from the three high days was 2300 per 100 ml. and the coliform density from the three low days was 43 per 100 ml. The difference in the illness rate from the second beach, where the mean coliform density from the three high days was 730 per 100 ml, was not great enough to rule out the possibility that it may have resulted from chance. The second study was conducted at a beach in Dayton, Kentucky, on the Ohio River near Cincinnati, Ohio. In this study an excess of gastrointestinal illness in swimmers was observed when the median coliform density was about 2300 per 100 ml. The third study conducted by the PHS was at two marine beaches on Long Island Sound, one at New Rochelle, New York, and the other at Mamaroneck, New York. The median density of coliforms in the New Rochelle Beach waters during the study was 610 per 100 ml, while at the Mamaroneck beach the median coliform density was 253 per 100 ml. No excess gastrointestinal illness in swimmers was observed at either beach.

The results from these studies suggested that swimming-associated gastroenteritis was not evident until the quality of the bathing water exceeded a median coliform density of at least 2300 per 100 ml. The results indicate that most of the states participating in the Garber survey in 1963 had set very conservative water quality standards.

In 1968 the Federal Water Pollution Control Administration commissioned the National Technical Advisory Committee (NTAC, 1968) to examine water quality criteria for a number of water use issues, including the relationship between swimming-associated illness and water quality. The Committee discussed the shortcomings of coliforms as an indicator of the presence of fecal contamination and declared that fecal coliforms were a much more specific indicator of fecal contamination. A study conducted on the Ohio River showed that fecal coliforms comprised about 18 % of the coliforms. This ratio was applied to the coliform density observed in the PHS studies where swimming-associated health effects were detected when the coliform density was about 2300 coliforms per 100 ml. The resulting equivalent fecal coliform density was approximately 400 fecal coliforms per 100 ml. The Committee suggested that a safety factor should be included in the new guideline, such that the water quality would be better than that at which a gastrointestinal health effect occurred, i.e., 400 fecal coliforms per 100 ml. The recommendation of the Committee was:

> Fecal coliforms should be used as the indicator organism for evaluating the microbiological suitability of recreational waters. As determined by multiple-tube fermentation or membrane filter procedures and based on a minimum of not less than five samples taken over not more than a 30-day period, the fecal coliform content of primary contact recreation waters shall not exceed a log mean of 200/100 ml, nor shall more than 10 percent of total samples during any 30-day period exceed 400/100 ml.

The NTAC recommendation was, in part, based on two unsupported assumptions. The first was that the fecal coliform portion of the coliform population was constant in all waterbodies of the United States, including marine waters. The data used by the NTAC had been obtained in the early 1960s from water samples collected from the Ohio River, which at that time received treated and untreated fecal wastes from many sources, not all of which were necessarily representative of sources across the United States. The second assumption, and possibly the more significant of the two, was the belief that by halving the indicator density at which a detectable health effect occurred, namely from 400 to 200 fecal coliforms, a zero risk would result. This assumption is considered unreasonable because the relationship between swimming-associated illness and water quality cannot be calculated from a single point (2300 coliforms per 100 ml). Both of these assumptions would have significant effects on later efforts to develop criteria and guidelines for the quality of recreational waters.

The potential for gastrointestinal illness to be associated with exposure to natural bathing waters was considered again in 1972 by the US Environmental

Protection Agency (EPA). Two series of studies were conducted, one at marine beaches (Cabelli, 1983) and the other at freshwater beaches (Dufour, 1984). The marine beaches were located in New York City at Coney Island and Rockaways beaches, in Boston at Revere and Nahant beaches, and at Lake Pontchartrain in the New Orleans area. The studies were conducted over a six-year period. Analysis of the data from the three locations shows that there was a linear relationship between the rate of swimming-associated gastroenteritis and the quality of the water measured with enterococci using the membrane filter technique but not so for *E. coli*. The swimming-associated illness rates ranged from zero per 1000 swimmers to 28 per 1000 swimmers with a mean rate of 15 illnesses per 1000. The water quality levels ranged from 3 enterococci per 100 ml to about 500 enterococci per 100 ml.

The freshwater studies were conducted at locations in Erie, Pennsylvania, on Lake Erie and at Keystone Lake about 60 miles east of Tulsa, Oklahoma. The studies took place in 1979, 1980, and 1982. The average rate of swimming-associated gastroenteritis was about 6 per 1000 swimmers with a high rate of 14 per 1000 swimmers and a low rate of zero per 1000. The mean water quality level, as measured with *E. coli*, was 72 per 100 ml and the mean level for enterococci was about 20 per 100 ml. The enterococci densities ranged from 6 to 80 per 100 ml. *E. coli* densities ranged from 18 to 250 per 100 ml. There were no significant illness correlations with other fecal indicators used in these studies.

In 1986, new criteria for recreational waters were recommended by the EPA, based on data collected during the above marine and freshwater studies (Federal Register, 1986). The recommendations were, in marine waters:

A geometric mean of 5 samples taken at equal time intervals over a 30-day period shall not exceed 35 enterococci per 100 ml.

and in freshwaters:

A geometric mean of 5 samples taken at equal intervals over a 30-day period shall not exceed 126 *E. coli* per 100 ml, and

A geometric mean of 5 samples taken at equal time intervals over a 30-day period shall not exceed 33 enterococci per 100 ml.

The upper threshold limit of enterococci in marine waters represents an acceptable risk of 19 swimming-associated illnesses per 1000 swimmers, while the upper limit for *E. coli* or enterococci in freshwaters represents an acceptable risk level of 8 gastrointestinal illnesses per 1000 swimmers. The basis for these acceptable risk levels is found in the EPA's desire to develop risk levels no more or less stringent than what was accepted under the fecal coliform limit of 200 fecal coliform per 100 ml. Although it was commonly believed that the 200 fecal coliform limit represented a zero risk, the EPA studies cited above clearly showed that this was not true.

The risk level for marine waters was based on three known values, all obtained from data collected during the epidemiological studies at marine bathing beaches They included the mean of the enterococci values per 100 ml from all of the beaches studied, a similar mean value for fecal coliforms and the recommended fecal coliform criteria of 200 per 100 ml. These values were used in the equation $A = (C \times B)/D$, where A is the unknown value of the criterion for enterococci, B is 29, the overall mean of enterococci, C is 200, the criterion for fecal coliform, and D is 166, the mean of fecal coliform from the marine beach epidemiological studies. Solving the equation gives 33 enterococci per 100 ml, the currently recommended criterion for marine waters; similar calculations were made for the freshwater criterion using a fecal coliform mean of 115 per 100 ml and an enterococci mean of 20 per 100 ml or an *E. coli* mean of 72 per 100 ml. The gastrointestinal illness rates of 8 and 19 per 1000 swimmers can be calculated using the regression equations found in EPA (1986). This ambient water quality criteria document also indicated single-sample maximum (SSM) values that might be used by states for other less frequent exposures to water than full body contact at designated beaches. These SSM values are are based upon the upper 75th percentile values for designated beach waters. The numbers have been established for both enterococci (62 per 100 ml) and *E. coli* (235 per 100 ml) in fresh water and enterococci (104 per 100 ml) in marine waters. The SSM values were set at the 82th percentile level for moderate full body contact recreation, at the 90th percentile value for lightly used full body contact and at the 95th percentile level for infrequently used full body contact. The SSM is viewed by EPA as a useful measure for making beach notification and closure decisions, especially in situations where sampling is infrequent, e.g., once a week. At some point in time many beach managers began monitoring their waters on a daily basis and the SSM allowed them to make decisions based on the daily sample.

The methods chosen for measuring fecal indicators are as important as the selection of indicators used for measuring water quality, since the methods govern how accurate, precise, and specific the water quality measurements are. The goal in 1915 of water microbiologists was to be able to quantify the presence of fecal contamination using microbes that were constantly associated with feces. The technology for growing fecal associated organisms was very crude and, in the case of coliforms, the most effective means of detecting these organisms was in a tube of liquid culture medium containing lactose. The lactose was readily fermented by coliforms and in the process produced large amounts of gas. The gas produced by coliforms in the liquid medium was captured in a second small tube inverted in the culture tube. The captured gas was easy to visualize and indicated that the tube contained growing coliforms. McCrady (1915) laid the foundation for the application of probability theory to the bacteriological examination of water by the fermentation tube method. His method, which was later improved by others, was the key to estimating the density of coliforms in water. This method for estimating the number of coliforms is called the most probable number (MPN) method and has been in use for almost 100 years. The method does have some drawbacks. It is very imprecise, it somewhat overestimates the true density of conforms in the water, and it cannot easily accommodate large volumes of sample.

In the early 1950s another method for measuring coliforms was introduced to water microbiologists by Goetz and Tsuneishi (1951). It made use of cellulose membranes, through which water samples could be passed. The membranes had a porosity which retained particles the size of bacteria. After filtration the membrane filters were then placed on growth medium, where, after a suitable incubation period, colonies of bacteria grew on the membrane. The colonies were easily visualized and the numbers of colonies per test volume of filtered sample could be calculated. Clark *et al.* (1951) described the use of membrane filtration (MF) for measuring coliforms. Some of the advantages of MF method, relative to the MPN method, were that it was more precise, it could accommodate larger water samples, the colony count results were more reproducible, and the assay system required less time to perform. MF disadvantages are that turbid waters, for example those contaminated with clay or algae, can foul the membranes and preclude the use of the test, and competition from high densities of non-target bacterial colonies can influence test results.

The PHS epidemiological studies were conducted using the MPN procedure for measuring water quality (Stevenson, 1953). The EPA epidemiological studies in the 1970s and early 1980s used MF methods to measure water quality (US EPA, 1986). Prior to 1960 all states used the MPN method to measure their water quality. In the NTAC (1968) report both the MPN and MF were recommended for measuring recreational water quality. The 1986 water quality criteria for recreational waters recommended only the MF method (Federal Register, 1986). It is not always clear in the beach survey literature what method is used by states to determine the quality of their waters, although most states do use the MF test. However, several states continue to use the MPN procedure.

In 1988, the EPA surveyed the 50 states to determine what standards they had in place (US EPA, 1988). Forty-six of the states had adopted the fecal coliform as the indicator of choice. Three states were using enterococci for marine waters and one state had selected *E. coli* for measuring the quality of freshwaters. Forty-three of the states used a geometric mean of 200 per 100 ml as a control limit and three used a single-sample fecal coliform limit of 200 per 100 ml. Four states used a median as the controlling limit. Three states used a control limit of fecal coliform less than the 200 per 100 ml limit. Three states used total coliforms as a second indicator, usually for freshwaters. Significant differences in the 1988 survey, relative to the 1963 survey conducted by Garber, were the indicator of choice and the calculation of the upper control limit value. In 1963, all 50 states used coliforms to measure water quality, whereas in 1988, 46 used fecal coliforms, the indicator recommended by NTAC (1968) and US EPA (1976). In 1988, 43 states had adopted the geometric mean to calculate the water quality value. In 1963, over half of the states used the arithmetic mean. The almost total conversion to the use of the geometric mean was more likely due to its recommendation by the NTAC and the U.S. EPA, although it appears that no formal discussion of why it should be used took place.

In 1992, the EPA published a report summarizing the water quality standards for bacteria that were then in force for each of the 50 states (US EPA, 1992). The listing indicated that for freshwaters, used for primary contact recreation, 43 states used fecal coliforms and the control limits ranged from 22 to 200 per 100 ml. Several states used values in the 2000 per 100 ml range. Seven states used either enterococci or *E. coli*, usually with an upper limit of 33 per 100 ml or 126 per 100 ml, respectively. Eighteen states developed or used marine water quality standards for fecal coliforms with upper control limits of 14 to 200 per 100 ml. Four states adopted enterococci or *E. coli* standards with upper limits of 7 to 35 for the former and 126 for the latter.

In 2003, the EPA updated the list of individual state bacterial water quality standards that were in place (US EPA, 2003). There was not much change in the distribution of the types of indicators used to measure water quality. In freshwaters there was a decrease from 43 to 40 in the number of states using a fecal coliform standard and for marine waters there was a decrease of six, from 18 states to 12, states that used fecal coliforms, and an increase in the use of enterococci for *E. coli* from four states to 11 states. The control values for enterococci and *E. coli* tended to follow the EPA recommended values; however, one or two states chose to put standards in place that were more stringent than the values recommended by the EPA. A more recent listing of state standards is not available, but it is likely that more states, especially coastal states, will adopt *E. coli* or enterococci as a result of the passage of the Beaches Environmental Assessment and Coastal Health Act in 2000, which encouraged states to adopt new standards based on the 1986 recommendations suggested by the EPA. There are some interesting findings in the latest listings of microbial standards used by states. First, there is still a great reluctance by many states to change their fecal coliform standards to ones that are more contemporary, for example, enterococci or *E. coli*. Second, there is some variability in the selection of upper control limits, indicating a degree of uncertainty by some states relative to which acceptable risk level is appropriate for primary contact recreational waters.

The 1986 based criteria, although they appear to be quite different from early water quality criteria, have threads that go back to the 1000 coliform per 100 ml standards in use before 1968. The PHS studies observed a detectable health effect when the coliform density was about 2300 per 100 ml (Stevenson, 1953). The NTAC (1968) translated the coliform level to 400 fecal coliforms based on a coliform to fecal coliform ratio, and then halved that number to 200 fecal coliforms (as a safety factor), which, in theory, was the equivalent of 1000 coliforms. In 1986 the EPA translated the 200 fecal coliform recommendation to *E. coli* and enterococci in order to maintain a risk level generally accepted prior to that time. What we have then, is 126 *E. coli* and 33–35 enterococci being equivalent to 200 fecal coliforms, and by extension equivalent to 1000 coliforms, a water quality standard in place in many states before any health studies were conducted. The reluctance to change health risk levels, and the associated bacterial densities they are related to, has made the development of recommendations or standards for recreational waters difficult to reconcile with scientific findings. The different approaches used over the years to develop standards for recommended criteria has in part been due to the way in which data are interpreted and that interpretation has relied frequently on the contemporary statistical tools available at the time. Issues that have been debated historically are still discussed today. Notable among those issues is whether to use the arithmetic mean or the geometric mean to characterize fecal contaminated waters using microbial measures. Current practice favors the use of the geometric mean for characterizing the quality of recreational waters. Although this subject has surfaced from time to time, no thorough discussion of the issue has taken place. Another issue that is quite obvious, if one reviews the development of criteria and the setting of standards in past years, is acceptable risk. How to select an acceptable risk level has been a much debated subject in the last half-century and one that has never come to closure. For some zero risk is the only acceptable risk, while for others risks other than zero are quite acceptable. This problem is due to the fact that there is no straightforward process leading to a judgment about where an acceptable risk level should be set. These issues and others will be discussed, in part, in the chapters that follow, in an attempt to clarify and define the statistics and procedures for evaluating microbiological and health data collected during the conduct of epidemiological studies and the design of plans for monitoring beach waters.

References

American Public Health Association (1922) Report of the Committee on Bathing Places. *American Journal of Public Health*, 12: 121–123.

American. Public Health Association (1924) Report of the Committee on Bathing Places. *American Journal of Public Health*, 14: 597–602.

American Public Health Association (1933) Swimming pools and bathing places. *American Journal of Public Health*, 23: 40–49.

American Public Health Association (1936) Bathing places. *American Journal of Public Health*, 26: 209–219.

American Public Health Association (1940) Look forward in the bathing place sanitation field. *American Journal of Public Health*, 30: 50–51.

American Public Health Association (1957) *Recommended Practice for Design, Equipment and Operation of Swimming Pools and Other Public Bathing Places*. American Public Health Association, Washington D.C.

Bureau of Sanitary Engineering (1943) Report on a pollution survey of Santa Monica Bay beaches in 1942. Report to California State Board of Public Health, Sacramento.

Cabelli, V.J. (1983) Health effects criteria for marine waters. EPA-600/1-80-031. US Environmental Protection Agency, Research Triangle Park, NC.

Clark, H.A., Geldreich, E.E., Jeter, H.L., and Kabler, P.W. (1951) The membrane filter in sanitary bacteriology. *Public Health Reports*, 66: 951–977.

Dufour, A.P. (1984) Health effects criteria for fresh recreational waters. EPA-600/1-84-004. US Environmental Protection Agency, Research Triangle Park, NC.

Federal Register (1986) Bacteriological ambient water quality criteria. *Federal Register*, 51(45): 8012–8016.

Goetz, A., and Tsuneishi, N. (1951) Application of molecular filter membranes to the bacteriological analysis of water. *Journal of the American Water Works Association*, 43: 943–984.

Kehr, R.W., and Butterfield, C.T. (1943) Notes on the relation between coliform and enteric pathogens. *Public Health Reports*, 58: 589–607.

McCrady, M.H. (1915) The numerical interpretation of fermentation-tube results. *Journal of Infectious Diseases*, 17: 183–212.

National Technical Advisory Committee (NTAC) (1968) *Water Quality Criteria*. Washington, DC: Federal Water Pollution Control Administration, Department of the Interior.

Orland, H.P. (ed.) (1965) *Standard Methods for the Examination of Water and Wastewater*, 12th edn. American Public Health Association, New York.

Pomeroy, R. (1955) Statistical analysis of coliform data – a discussion. *Sewage and Industrial Wastes*, 27: 1299–1301.

Parkhurst, D.F. (1998) Arithmetic versus geometric means for environmental data. *Environmental Science and Technology*, Feb. 1, 1998: 92A–98A

Scott, W.J. (1932) Survey of Connecticut's shore bathing waters. *American Journal of Public Health*, 22: 316–321.

Senn, C.L. (1963) Coliform standards for recreational waters. Progress Report of the Public Health Activities Committee. *Journal of the Sanitary Engineering Division, Proceedings of the American Society of Civil Engineers*, 89(SA4): 57–94.

Stevenson, A.H. (1953) Studies of bathing water quality and health. *American Journal of Public Health*, 43: 529–538.

Streeter, H.W. (1951) Bacterial quality objectives for the Ohio River, pp. 24–25. Ohio River Valley Water Sanitation Commission, Cincinnati, OH.

Thomas, H.A. (1955) Statistical analysis of coliform data. *Sewage and Industrial Wastes*, 27: 212–222.

US Environmental Protection Agency (1976) Quality criteria for water, 1976. US Environmental Protection Agency, Washington, DC.

US Environmental Protection Agency (1986) Quality criteria for water – 1986. EPA 440/5-86-001. US Environmental Protection Agency, Office of Water, Washington, DC.

US Environmental Protection Agency (1988) Water quality standards criteria summaries: A compilation of state/federal criteria. EPA 440/5-88-007. US Environmental Protection Agency, Office of Water, Washington DC.

US Environmental Protection Agency (1992) Summary of state water quality recreational standards for bacteria. US Environmental Protection Agency, Office of Water Policy Analysis, Washington, DC.

US Environmental Protection Agency (2003) Bacterial water quality standards for recreational waters (freshwater and marine waters). Status Report. EPA 823-R-403-08. US Environmental Protection Agency, Washington, DC.

2

A management context for the statistical design of recreational contact water quality monitoring programs

Stephen B. Weisberg
Southern California Coastal Water Research Project, Costa Mesa, CA, USA

When considering the statistical design for a beach monitoring program, it is important to recognize that a single sampling design will not be appropriate for all management applications. Many factors influence sampling design selection, the most important of which is the monitoring objective. Typically, precision of sampling estimates is a driving force in selecting a sampling design; however, precision is only one of the uncertainties associated with the manner in which beach water quality data are used. The amount of effort expended on enhancing precision should be guided by the nature of the decision that is being made and the context of other uncertainties associated with the decision. This chapter describes the most common applications of beach monitoring data and sources of uncertainty inherent in these decision frameworks.

Statistical Framework for Recreational Water Quality Criteria and Monitoring   Edited by Larry J. Wymer
© 2007 John Wiley & Sons, Ltd

2.1   Health warning decisions

The most common use of beach water quality data is for public health warnings in which local officials use monitoring data in deciding whether to issue an advisory that swimmers stay out of the water. The nature of the warning varies from a simple advisory that poor water quality has been observed to a beach closure that prohibits swimmers from entering the water. Advisories are generally viewed as a right to know, with emphasis placed on minimizing Type I error. Closures, on the other hand, have a more direct economic impact and priority is placed on minimizing Type II error.

In making closure decisions, local health officials rely on three types of assumptions. The first is that the routine monitoring data provide an accurate assessment of water quality conditions at the time of sampling. Most of this book addresses the ways in which the allocation of sampling intensity in space and time is weighed against the precision of laboratory measurements to develop an appropriate sampling design. However, no matter how precise this estimate is, these estimates are only one leg of a three-legged decision stool.

The second leg is an assumption about how well measurement of one day's condition predicts the next day's condition. Present laboratory enumeration methods require approximately 24 hours for processing; consequently, measurement on any given day is used to predict the likelihood of a condition that will occur on the subsequent day. The need for precision about existing conditions must be tempered against the imprecision of the extrapolation to the latter day (Leecaster and Weisberg, 2001; Boehm *et al.*, this volume). Accuracy of the prediction is partially dependent on the type of beach being sampled. Beaches in enclosed waterbodies that have limited circulation, or beaches near a known continuous contamination source, are more likely than open coastal beaches to sustain water quality conditions for several days. Residence time for a parcel of water on open coastal beaches can be less than one hour because they are subjected to circulation by ocean currents (Boehm *et al.*, 2002). Developing great precision about present conditions for beaches that change condition rapidly would be a poor investment of sampling resources. Because of high circulation associated with open coastal beaches, public health officials in these areas often require several days of high bacterial counts before enacting a closure to ensure that the problem is not transient.

The third assumption in the health warning decision is that there exists a good relationship between the indicators being measured and potential health risks. The use of fecal indicator bacteria is premised on its relationship to health risk established through epidemiological studies (Wade *et al.*, 2003). However, these studies have been primarily conducted at beaches where sewage effluent is the primary source of contamination and the health risk relationship may not be as robust as the relationship in locations where contamination is primarily derived from nonhuman fecal material. As a result, several states, such as California and Hawaii, incorporate multiple indicators in their sampling designs and look for concordance among several bacterial indicators as an indication of a likely human fecal source. The National Research Council (2004) suggests that multiple indicators be implemented in a tiered adaptive sampling design, with the least expensive and most rapid methods used for screening, reserving more expensive, human-specific measures for use in confirmation of health risk. Presently, many health officers use visual surveys for sewage leaks or knowledge of likely local sources (e.g. nearby septic systems) as confirmatory evidence of health risk. The robustness of the epidemiological relationships and potential benefits of incorporating multiple indicators in a sampling design are discussed further in Chapter 3.

The desired relation to health risk affects selection of sampling methods, as there is a preference for replicating methods used in the epidemiological studies that generated the standards on which the decisions are based. For example, epidemiological relationships based on samples collected at noon in ankle-deep water may not apply well to samples collected in waist-deep water in the early morning. Similarly, a standard based on culture-based laboratory methods may not be perfectly applicable to samples processed with new molecular-based methods (Noble and Weisberg, 2005).

2.2 Compliance monitoring

While monitoring for health warning systems is the most widely publicized use of beach water quality data, monitoring in many parts of the USA is associated with compliance determinations for treated wastewater and stormwater discharge systems (Schiff *et al.*, 2002). Most treated wastewater dischargers and stormwater programs have bacterial monitoring components. At sites where disinfection is part of water treatment processes, monitoring may be limited to end of the discharge pipe, but in many cases compliance monitoring also includes a component for monitoring at nearby beaches.

Compliance monitoring differs from health warning systems in that it is largely retrospective, with a goal of determining whether a problem occurred in the past, rather than whether it is likely to occur in the future. Short-term management decisions driven by this type of monitoring are typically about whether a fine should be levied for exceedance of water quality standards. In the longer term, these data are used to determine whether upgrades to treatment processes are required.

From a design perspective, the question is primarily one of estimating the percentage of time a compliance standard is exceeded. Thus, there is less need to be precise in the estimate for any given day. This may lead to a reduced need for spatial replication and an increased need for adaptive designs that involve more extensive monitoring when conditions suggest a violation is likely to occur (e.g. in-plant monitoring of other parameters suggests that treatment processes are not functioning properly).

Discharge compliance monitoring also differs from health warning monitoring in that there is a need to associate exceedances with a particular source. Most beaches, even those adjacent to a wastewater outfall, are subject to contamination from many sources, including land-based runoff, shorebirds, and humans (particularly young children) bathing at the beach. Before fines are levied for exceedances, there is a need to ensure that the discharger is the source of the water quality problem. This is typically implemented through analysis of spatial gradients from the source, and sampling designs must be particularly robust for this type of assessment. In a smaller number of cases, assessment is achieved through the measurement of additional indicators incorporated in an adaptive format that allows for source attribution.

2.3   Trend assessments

A third type of monitoring objective is trend assessments, in which the goal is to assess the ways in which water quality conditions are changing over longer periods of time. This is sometimes done to assess whether continuing development is leading to an increased likelihood of future water quality problems. Often this is done to assess whether management actions are successful in resolving a past problem. From a design perspective, there is less need for certainty about conditions on a particular day and more need for temporal coverage.

Method sensitivity becomes a more important design issue when the goal is trend monitoring. Discharge compliance and health warning monitoring data are typically assessed relative to a standard, requiring little need for sensitivity or precision below the standard. In contrast, trend monitoring often focuses on temporal patterns in bacterial concentrations below the standard and may require the use of more sensitive laboratory methods that minimize the number of values below (or above) detection limits. From a design perspective, the cost of more expensive laboratory methods, or processing of multiple dilutions to ensure quantification over a larger range, must be weighed against the costs of additional spatial and temporal allocation of sampling effort.

Method consistency also becomes more important in trend monitoring. New methods that lower cost or improve precision may be developed over time; however, such methods may not be preferable in a trend assessment with an increased emphasis on the avoidance of confounding method bias with temporal trends. Intercalibration tests, though, provide a means for quantifying method bias and placing results in context of anticipated temporal changes (Griffith *et al.*, 2006).

2.4   Summary

This book focuses on the use of statistical methods in beach water quality monitoring and appropriately emphasizes spatial and temporal allocation of sampling efforts to increase precision of bacterial concentration estimates for an individual day. However, when applying these principles to enhance precision, it is important to remember that a single sampling design will not be appropriate to all applications. Other factors associated with uses of the data should also influence the selection of sampling design.

References

Boehm, A.B., Grant, S.B., Kim, J.H., Mowbray, S.L., McGee, C.D., Clark, C.D., Foley, D.M., and Wellman, D.E. (2002) Decadal and shorter period variability and surf zone water quality at Huntington Beach, California. *Environmental Science and Technology*, 36: 3885–3892.

Griffith, J.F., Aumand, L.A., Lee, I.M., McGee, C.D., Othman, L.L., Ritter, K.J., Walker, K.O., and Weisberg, S.B. (2006) Comparison and verification of bacterial water quality indicator measurement methods using ambient coastal water samples. *Environmental Monitoring and Assessment*, 116: 335–344.

Leecaster, M.K., and Weisberg, S.B. (2001) Effect of sampling frequency on shoreline microbiology assessments. *Marine Pollution Bulletin*, 42: 1150–1154.

National Research Council (2004) *Indicators for Waterborne Pathogens*. National Academies Press, Washington, DC.

Noble, R.T. and Weisberg, S.B. (2005) A review of technologies being developed for rapid detection of bacteria in recreational waters. *Journal of Water and Health*, 3: 381–392.

Schiff, K.C., Weisberg, S.B., and Raco-Rands, V.E. (2002) Inventory of ocean monitoring in the Southern California Bight. *Environmental Management*, 29: 871–876.

Wade, T.J., Pai, N., Eisenberg, J.N.S., and Colford, J.M., Jr. (2003) Do US EPA water guidelines for recreational waters prevent gastrointestinal illness? A systematic review and meta-analysis. *Environmental Health Perspectives*, 111: 1102–1109.

3

Conceptual bases for relating illness risk to indicator concentrations

D.F. Parkhurst, G.F. Craun, and J.A. Soller

Environmental Science Research Center, School of Public and Environmental Affairs Indiana University, Bloomington, IN, USA, Gunther F. Craun & Associates, Staunton, VA, USA, and Soller Environmental, Berkeley, CA, USA

3.1 Introduction

Public health officials have long relied on the indicator organism approach to evaluate water quality. Historically, these indicators were used to assess the degree to which water was polluted by fecal contamination from humans and other warm-blooded animals. They provided a measure of the opportunity or possibility for transmitting a waterborne disease that might result from fecal contamination. The indicators were not expected to have a direct relationship to the numbers of pathogens present in the water, and they were not used to determine the actual risk of contracting a specific waterborne disease. For example, bacterial indicators (members of the coliform group) have been used for over 75 years to assess the microbiological quality of drinking water. Although coliform bacteria can be used

Statistical Framework for Recreational Water Quality Criteria and Monitoring Edited by Larry J. Wymer
© 2007 John Wiley & Sons, Ltd

to assess the potential for bacterial pathogens in drinking water, studies have shown that they are inadequate indicators for protozoan and viral waterborne pathogens and that the total coliform rule (US EPA, 1975, 1989, 1990) is inadequate to identify water systems that are vulnerable to an outbreak (McCabe, 1977; Batik *et al.*, 1983; Craun *et al.*, 1997; Nwachuku *et al.*, 2002).

Desirable characteristics of water quality indicators have been discussed by many authors (National Research Council (NRC), 1977; McFeters *et al.*, 1978; Allen and Geldreich, 1978; Sobsey *et al.*, 1995; Toranzos and McFeters, 1997) and it is generally agreed that a good indicator should:

- be present whenever pathogens are potentially present, *i.e.,* be present in fecally contaminated waters at easily detectable levels, yet be absent in uncontaminated waters;

- occur in much greater numbers than pathogens;

- be more resistant to degradation in the aqueous environment than pathogens;

- be easily detectable in fecally contaminated water using simple methods;

- be removed and inactivated by water treatment processes similarly to, or less easily than, disease-causing pathogens;

- not multiply in the environment.

McCabe (1977) noted the lack of epidemiologic investigations relating water quality indicators to health risks, and it is now recognized that indicators should also allow predicting human health risk. For example, recreational water quality guidelines in the United States have been developed on the basis of epidemiologic studies that associated gastrointestinal illness with the presence of enterococci or *Escherichia coli*. The EPA guideline for freshwaters is a monthly geometric mean of no more than 33 enterococci per 100 ml, or no more than 126 *E. coli* per 100 ml for marine waters; the monthly geometric mean should be no more than 35 enterococci per 100 ml (US EPA, 1986, 2002b; Dufour *et al.*, 1984; Cabelli, 1983; Cabelli *et al.*, 1982). These or other indicator concentrations are used by state and local governments to determine when to close beaches or post warning signs to alert potential bathers of poor water quality.

The EPA is currently evaluating the relationship between various health risks (gastroenteritis, skin rash, and infections of the respiratory system, throat, eyes, and ears) and concentrations of enterococci and/or *Bacteroides* in water as measured by quantitative polymerase chain reaction (QPCR) analyses. QPCR analyses can provide results within two hours, much faster than analytical methods currently used for enterococci or *E. coli*.

The general idea of using an indicator organism to estimate or predict risk of illness in swimmers begins with the supposition that the concentration of an indicator in the water increases with the amount of fecal material in the water.

Fecal material *per se* does not make people ill; pathogens in the fecal material may, depending upon host characteristics and dose. Thus, a good indicator should correlate with or measure human sewage and/or animal waste, which in turn will contain different mixes and/or quantities of pathogens at any one time. Furthermore, the indicator should be predictive of human illness risks in a general way.

Sources of pathogens in recreational water include domestic sewage discharges from wastewater treatment plants, on-site systems such as septic tanks, and non-point sources of animal or human waste. Although there is a supposition that indicator levels increase with pathogen levels, this may not always be the case. The relationship between fecal indicators and human enteric pathogens varies with the type of contamination sources and with season. Most enteric bacteria and protozoan parasites are harbored by a variety of animals; most enteric viruses of concern to humans are harbored only by humans (Cotruvo *et al.*, 2004). The number and variety of pathogens excreted by populations into sewage that is discharged (after treatment) at point sources will vary from place to place and from time to time at any given location, depending upon infection and illness of the population contributing to the sewage.

Non-point sources can also vary by season and location. Furthermore, pathogens and indicators may have different die-off rates through sewage treatment, in the soil environment, and in receiving waters (Allwood *et al.*, 2003). Pathogens may also have different transport properties in water. In addition, the potency and virulence of pathogens can vary. Some cause illness only at relatively high exposure levels, but others at low exposure levels (Geldreich, 1991). For example, the infectious dose to produce clinical symptoms in 50 % of individuals can range from 10^2 organisms or fewer (e.g., *Shigella*, enterovirius, *Giardia*, *Cryptosporidium*) to 10^6–10^9 organisms (e.g., *Salmonella*, *Campylobacter*, *Yersinia*). Human volunteer studies also show that the dose–illness relationships vary from one pathogen to another; see Section 3.3 below. In addition, the quantity of pathogens excreted by infected persons may range from 10^5 *Giardia* cysts or *Yersinia* bacteria per gram of feces to 10^8 toxigenic *E. coli* per gram of feces (Geldreich, 1991).

When considering water quality indicators that might predict health risks, one should remember that the risk of contracting waterborne disease cannot necessarily be evaluated simply by determining the presence of a pathogen in the water. The mere detection of the microorganism provides inadequate evidence about its pathogenicity, virulence, and potency. The capability of a microorganism to cause disease in a specific population by a specific exposure route (e.g., recreational water) depends on not only the dose but also the population characteristics and susceptibility of the exposed individuals. In addition, the particular microorganism detected may lack an essential characteristic required to cause illness. Other considerations include the prevailing environmental conditions that may influence the fate and transport of the pathogen and its hosts, and the role of other animal reservoirs, vectors, and vehicles of transmission. In addition, while some analytical methods can culture only live microorganisms from a water sample, other methods detect microorganism activities (ATP or oxidation-reduction potential), antigens, other proteins, and genes. Thus, detection methods for pathogens in water do not always provide evidence of infectivity.

Microorganisms found in water may have both pathogenic and non-pathogenic strains. This phenomenon has been documented for bacteria such as *E. coli*, *Aeromonas hydrophila*, and *Yersinia enterocolitica* (Agarwal *et al.*, 1998; Bottone, 1999; Isonhood and Drake, 2002; Ryan, 1994; Schaechter, 1998; Stephen, 2001). Various isolates of *Cryptosporidium* are infectious (virulent) to varying degrees in humans (Okhuysen *et al.*, 1999, 2002; Okhuysen and Chappell, 2002; Widmer *et al.*, 1998). Xiao *et al.* (2000) reported that *Cryptosporidium parvum* is made up of a number of genotypes, each with different infectivity or virulence for human populations. Using genetic methods, Morgan *et al.* (1999) identified eight different species of *Cryptosporidium* and seven genotypes of *C. parvum*. An eighth genotype was identified by Sulaiman *et al.* (1999, 2000). *C. parvum* genotype 1 may infect only humans; other genotypes may infect various mammals. Hepatitis E virus, *E. coli*, *Shigella*, *Legionella*, *Salmonella*, and *Yersinia* are among other waterborne pathogens for which species, strains, serotypes, or genotypes differ with respect to risks and severity of human disease (Adam, 2000).

One important factor in determining illness among swimmers is the human response to infection. If a person is exposed to an infectious waterborne agent through ingestion or inhalation, the person may become infected. However, for an infection to occur, (1) a microorganism must negotiate its way past a variety of barriers in the host and reach the site of susceptible cells in tissues of the target organ (e.g., intestinal tract, lung), (2) sufficient numbers of the microorganism must make contact with the cells of this tissue and must successfully multiply intra- or extracellularly (in or on the surface of the cells) of the susceptible tissue, and (3) the defenses of the cell, tissue, and organ system must be overcome. The agent may further multiply and spread to other cells and tissues. If the infectious agent is then to cause disease, (1) it must possess specific virulence factors, (2) those virulence factors must be expressed, (3) infection must occur at a site in the host where the disease phenomenon or pathology is expressed, and (4) host defense systems must be overcome. Thus, the severity of any resulting disease depends not only on the pathogen's potency and virulence but also on the susceptibility of the host. The illness resulting from infection may be inapparent (asymptomatic), mild, or severe. Waterborne outbreaks associated with recreational waters and reported in the United States have caused mild gastroenteritis as well as hospitalizations and deaths (Craun *et al.*, 2005).

For certain viruses, the ability to infect and cause illness in a human host depends primarily on the genetics of the host. Hosts that lack a cell surface receptor for virus infection are recalcitrant to infection (or innately immune), and therefore, for these hosts, the viruses are harmless. A good example of this is norovirus, for which susceptibility to infection and illness is dependent on the presence of a host cell receptor for the virus that is a blood group antigen (Hutson *et al.*, 2002; Marionneau *et al.*, 2002; Lindesmith *et al.*, 2003). Humans of certain blood groups lack the receptor and are completely resistant to infection.

Host susceptibility can vary both within a community and between communities. Persons with increased risk of disease and severity of disease include the very young and the elderly, pregnant women, undernourished individuals, and patients with compromised immunity due to diseases such as AIDS, and to medical interventions such as organ transplants and cancer treatment. Waterborne pathogens may have a greater impact on persons who are malnourished or already suffering from other disease. Possible protective immunity is also a factor in the relationships among indicators, pathogens, and waterborne risks. For example, seroepidemiological studies suggest that immunity is important when assessing waterborne *Cryptosporidium* risks (Craun *et al.*, 2004; Frost *et al.*, 2003, 2005). For local beaches with sewage discharges, persons contributing to the sewage contamination of the recreational water may also be those who use the beach most often. In this scenario, persons who come from outside the local area to swim may be more vulnerable to illness when exposed, due to their having lower immunity. However, not all waterborne pathogens confer protective immunity, and some immunity may be only short-lived (Cotruvo *et al.*, 2004).

For these reasons, it is important that epidemiologic studies of populations at various water recreational sites be conducted to provide the evidence for an association between specific levels of any proposed indicator(s) and human health risks. In evaluating the indicator–human health relationship, investigators should consider the characteristics of the indicators, their relationship to pathogen concentrations, sources of contamination, host population susceptibility, and other factors that might contribute to the variability and uncertainty of the indicator–health risk relationship. In this chapter, we discuss the rationale and limitations inherent in developing indicators that can predict illness risks.

3.2 Pathogens of public health concern in recreational waters

3.2.1 Waterborne pathogens

Waterborne infectious agents of intestinal origin discharged in human and animal feces fall into four broad groups: viruses, bacteria, protozoa, and helminths. A brief description of the characteristics of the various categories of microbial pathogens is provided below (Olivieri and Soller, 2001; Soller *et al.*, 2004). Symptoms associated with waterborne microbial infections are also briefly summarized by Feachem *et al.* (1983) and Mead *et al.* (1999). As noted previously, not all infected individuals exhibit symptoms of disease. The proportion of infections that result in symptomatic illness varies from pathogen to pathogen (Soller *et al.*, 2004).

Viral pathogens

Viruses are obligate intracellular parasites having no cell structure of their own, and are therefore incapable of replication outside a host organism. They range in size from approximately 0.025 to 0.350 μm, and thus can only be observed with an electron microscope. Over 140 types of known human enteric viruses are known. These viruses replicate in the human intestinal track and are shed in fecal material of infected individuals. The term 'enteric viruses' is applied to any viruses disseminated by the fecal route. They are further divided into several groups based on morphological, physical, chemical, and antigenic differences. The most commonly studied group in water is the enteroviruses, which include poliovirus, coxsackievirus, and echovirus, for example.

Viruses potentially found in raw wastewater include adenoviruses, rotavirus, reoviruses, enteroviruses, calicivirus, astrovirus, noroviruses, hepatitis A, and poliovirus (US EPA, 2002a). Waterborne viral infections can lead to a wide range of adverse health effects, including gastroenteritis, vomiting, diarrhea, meningitis, sore throat, or flu-like symptoms, depending on the specific pathogen. Hepatitis A infections can cause jaundice, fatigue, and fever in addition to some of the symptoms caused by infections of other viral pathogens.

Bacterial pathogens

Bacteria are microscopic, mostly unicellular organisms with a relatively simple cell structure lacking a cell nucleus, cytoskeleton, and organelles. They typically range in size from 0.2 to 10 μm and are the most abundant of all organisms. Fecal material contains many types of bacteria, many of which are harmless. Fecal bacteria colonize the human intestinal tract, and feces can contain up to 10^{12} bacteria per gram. As mentioned earlier, one group of intestinal bacteria, the coliform bacteria, has historically been used as an indicator of fecal contamination. Other important bacteria that are both pathogenic and transmittable by the waterborne route (e.g., *Salmonella*, *Campylobacter*, *Yersinia*, and *E. coli* O157:H7) may be present in the feces of humans or other animals. The only significant reservoir for *Shigella*, an important waterborne pathogen, is humans.

Bacterial pathogens known to be present at times in raw wastewater include: *Campylobacter* spp., *E. coli* O157:H7 and other pathogenic forms, *Vibrio cholerae*, *Pseudomonas aeruginosa*, *Salmonella typhi* and other species, *Shigella* spp., and *Yersinia* spp. *Salmonella* spp. are associated with salmonellosis, which is characterized by fever, abdominal cramping, and diarrhea. Diarrhea from *E. coli* O157:H7 may range from mild and non-bloody stools to stools that are virtually all blood; certain strains may also elaborate potent verotoxins causing hemolytic-uremic syndrome (Benenson, 1995). Infections of *Shigella* lead to bacterial dysentery or shigellosis, which is often characterized by bloody diarrhea, fever, and stomach cramping. Certain strains of *Shigella* (Benenson, 1995) can cause a reactive arthropathy (Reiter's syndrome) in persons who are genetically predisposed. *Yersinia enterocolitica* typically causes acute febrile diarrhea and lymphadenitis mimicking appendicitis; bloody diarrhea is seen in 10–30 % of infected children and joint pain is reported in half of infected adults (Benenson, 1995). Many infections with *Campylobacter jejuni* are asymptomatic; illness is characterized by diarrhea, abdominal pain, fever, nausea, and vomiting; reactive arthritis may also occur (Benenson, 1995). Cholera, in its severe form, is characterized by the sudden onset of profuse, watery diarrhea and in untreated cases by rapid dehydration, acidosis, circulatory collapse, and renal failure (Benenson, 1995).

Protozoan parasites

Most protozoan parasites produce cysts or oocysts that can survive outside their hosts under adverse environmental conditions. In general, protozoan parasitic cysts are larger than bacteria, ranging in size from 2 to 15 μm, and they survive for longer periods in the water environment. People with both symptomatic and non-symptomatic infections can excrete protozoan cysts or oocysts. Protozoan parasites are similar in nature to viruses in that they do not reproduce outside host organisms; however, many protozoa, including *Cryptosporidium* and *Giardia*, may be present in the feces of humans or other animals. Humans are the only known reservoir for *Entamoeba histolytic*, a waterborne agent of declining importance in the United States.

Protozoa such as *Cryptosporidium parvum*, *Cyclospora cayetanensis*, *Giardia intestinalis*, *Balantidium coli*, and *Toxoplasma gondii* may be found in raw wastewater as well as in surface water from non-point sources of contamination. *Cryptosporidium* causes gastroenteritis characterized by diarrhea, loose or watery stools, stomach cramps, upset stomach, and low-grade fever. Children and pregnant women are especially susceptible to dehydration from *Cryptosporidium*, and the disease may be life-threatening to those with weakened immune systems. *G. intestinalis* causes diarrhea, loose or watery stools, stomach cramps, and upset stomach. Additionally, *Giardia* infections may lead to weight loss and dehydration, particularly in children and pregnant women. *Giardia* is highly contagious, and generally lasts 2–3 weeks. Infections by other protozoa may lead to a wide range of ailments including flu-like symptoms, ulcers, abdominal pain, headache, fever, nausea, seizures, and amoebic encephalitis.

Helminthic parasites

Helminths of pathogenic importance include the roundworm and the hookworm. The ova, the infective stage of parasitic helminthes, are particularly resistant to environmental factors. Helminthic eggs and cysts tend to range in size from 5 to 150 μm and may be spherical or cylindrical. Helminthic infections tend to be transmitted via contaminated soil and food. Many helminthic infections in humans require an intermediate animal host. For the most part, helminths are not a major health problem in the United States, but they can be in other areas.

3.2.2 Waterborne pathogens identified in outbreaks reported in untreated recreational water

A recent review of the causes of outbreaks in recreational waters found that during the 30-year surveillance period from 1971 to 2000, 146 outbreaks were reported in untreated recreational waters (Craun *et al.*, 2005). Most frequently, exposure to contaminated recreational waters resulted in acute gastroenteritis and dermatitis. An etiologic agent was not identified in 21 % of the outbreaks. *Shigella* was the most frequently identified etiology of outbreaks (21 %) and illnesses (29 %) in untreated water. Other important agents included *Naegleria fowleri* (17 %), *E. coli* O157:H7 (9 %), *Schistosoma* causing swimmer's itch (9 %), norovirus (6 %), *Leptospira* (5 %), and *Giardia* (4 %). *Cryptosporidium* caused 3 % of the outbreaks. The remaining outbreaks (5 %) involved conjunctivitis and pharyngitis attributed to enterovirus or adenovirus, hepatitis A, and typhoid fever. Few studies other than those associated with outbreaks have been conducted to determine the etiological agents related to swimming-associated illness (World Health Organization (WHO), 1999).

3.2.3   Waterborne pathogens of primary public health concern in the United States

As noted above, a wide range of waterborne pathogens are found in wastewater and non-point sources of contamination of surface waters, and these pathogens can cause a wide range of illnesses. From a public health perspective, those pathogens responsible for the majority of illnesses in the USA should be of primary concern in assessment of recreational water risks and selection of appropriate indicators. For example, some pathogens may only be found in certain climates or areas, and although similar pathogens may be found in waters throughout the world, there can be differences in their concentrations and frequency of occurrence. In addition, the population characteristics (e.g., other sources of exposure, susceptibility) should be considered because they can affect the incidence of disease. Discussed here are potentially important pathogens in the USA.

The occurrence of food-related illnesses can help in identifying important pathogens. In characterizing food-related illness and death in the USA, Mead *et al.* (1999) estimated the annual total number of illnesses caused by known pathogens (adjusting for non-reporting of many illnesses). An estimated 38.6 million cases of illness occur annually in the United States, with 5.2 million cases from bacterial pathogens, 2.5 million from parasitic pathogens, and 30.9 million from viral pathogens.

Of the nearly 31 million illnesses caused by viruses in the USA annually, noroviruses are estimated to account for 23 million, 60 % of which are believed to be non-foodborne (e.g., day-care related, person-to-person transmission, and waterborne; Mead *et al.*, 1999). Although the first human noroviruses were discovered nearly 30 years ago, much of the epidemiological and biological character of these viruses is only now beginning to unfold. Investigation has been difficult due to a number of factors; the viruses cannot be amplified by *in vitro* cell culture or animal models, and electron microscopy is often not sensitive enough to detect them in stool samples.

Recent advances in molecular diagnostic techniques have highlighted the clinical and public health importance of noroviruses, their ability to cause infection via a number of transmission routes, and their considerable genetic diversity (Lopman *et al.*, 2002). Bennett *et al.* (1987) estimated that only 5 % of noroviruses are waterborne; however, viral waterborne outbreaks are likely to be underreported because improved technology for detection of viruses in stool and water samples is not widely applied (Yoder *et al.*, 2004). Data from the WHO (1999) suggest that noroviruses may be the important etiology of acute gastrointestinal illness observed in epidemiologic studies conducted at bathing beaches, and recent research indicates that noroviruses may be present in surface waters at concentrations that would be consistent with causing higher disease incidence than other viruses (Lodder and Husman, 2005).

Of the protozoa, *G. intestinalis* has been reported to cause 2 million illness cases per year in the USA, and *C. parvum* causes an estimated 300 000 illnesses each year. Mead *et al.* (1999) suggests that 90 % of these illnesses are transmitted by routes other than contaminated food.

In contrast to enteric viruses and those two protozoa, many of the important bacterial pathogens in the USA are thought to be primarily transmitted by the foodborne route (Mead *et al.*, 1999). Of those accounting for the non-foodborne illnesses, *Salmonella* spp., *Shigella* spp., and *Campylobacter* spp. together have been reported to account for approximately 4.3 million of the 5.2 million annual bacterial illness cases in the USA, of which approximately 95 %, 20 %, and 80 % respectively are foodborne (Mead *et al.*, 1999).

Because of the large number of illnesses caused by enteric viruses, and the high numbers of viruses excreted by infected individuals (Yates and Gerba, 1998), enteric viruses are potentially the most prevalent pathogens present in recreational waters impacted by wastewater treatment discharges. The potential importance of viral pathogens in waterborne transmission of disease through recreational activities is supported by research conducted over the last 20 years (Cabelli, 1983; Fankhauser *et al.*, 1998; Levine and Stephenson, 1990; Palmateer *et al.*, 1991; Sobsey *et al.*, 1995) and recent waterborne outbreaks (Maunula *et al.*, 2004, Hoebe *et al.*, 2004).

Based on those pathogens known to be present in wastewater and their estimated disease incidences, it is reasonable to infer that the pathogens of public health concern likely to be most important from exposure to recreational waters in the USA, and perhaps in other developed countries, include noroviruses, *C. parvum*, *G. intestinalis*, *Shigella* spp., and *E. coli* (O157:H7, verotoxigenic, and other serotypes). As noted previously, these pathogens are frequently identified as etiologic agents causing gastroenteritis outbreaks among bathers; less frequently identified agents included hepatitis A, *Salmonella typhi*, and adenovirus.

3.3 Dose–response curves for pathogens

Disease risks are sometimes modeled as functions of indicator organism concentrations, requiring a choice of functional forms. To the extent that pathogens provide the link between indicators and disease, the shapes of pathogen–disease risk curves may be relevant. Numerous studies have provided information about human responses to various pathogens, though not in swimming-related contexts. It is common to fit the sigmoidally shaped exponential and beta-Poisson models (e.g., Haas *et al.*, 1999; McBride *et al.*, 2002) to dose–response results (Figure 3.1), but curvature of the actual data appears to vary considerably among pathogens.

Figure 3.2 provides some examples of raw data, and also illustrates a tendency seen in these and other data for a given response rate (e.g., 10 %) to occur at small doses of viruses (measured as focus-forming units and plaque-forming units), intermediate numbers of protozoan pathogens, and large numbers of bacteria. These curves show that linear increases in dose do not necessarily result in linear increases in probability of infection, and that the shape of the observed dose–response relationships varies from one pathogen to another. For example, Figure 3.2a shows a nearly linear dose–response relationship, over a fairly wide range of infection risk, for rotavirus (Ward *et al.*, 1986); such linearity is not seen for all viruses, e.g. for echovirus 12 (Schiff *et al.*, 1984). Figure 3.2b shows the 'bent over' relationship for *C. parvum* (Dupont *et al.*, 1995), indicating relatively reduced gain in risk at higher doses, and Figure 3.2c presents the opposite curvature, in the risk range shown, from a study of *Salmonella enterica* by McCullough and Eisele (1951).

These curvatures are of interest because, lacking other information, one might suppose as a first approximation that pathogen numbers and indicator numbers would vary proportionately to one another. If that were the case, these different curvatures would indicate that the shapes of indicator–risk relationships could vary from time to time and beach to beach, depending on which pathogens were contributing the risks.

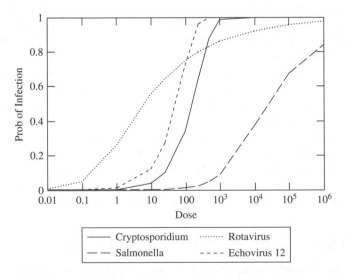

Figure 3.1   Dose–risk curves for selected pathogens.

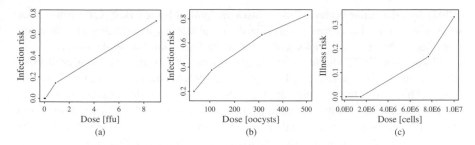

Figure 3.2 Example dose–risk curves for (a) viral (rotavirus; Ward *et al.*, 1986), (b) protozan (*C. parvum*; Dupont *et al.*, 1995), and (c) bacteria (S.*enterica* Var *meleagridis*, strain III; McCullough and Eiselk, 1951) pathogens. In each case, one or more high points has been omitted, to emphasize the forms of the relationships at lower risk levels of interest in protecting swimmers. ffu: focus-forming units.

3.4   Similarities and differences between pathogens and indicators

Indicator bacteria and pathogenic microbes may differ in several ways that could reduce the ability of the indicators to predict pathogen levels. Some possibilities are:

- die-off rates from sunlight, heat, salt, or other qualities of water;
- reductions of concentrations and viability from water treatment processes;
- loss to predation by zooplankton;
- sedimentation rates.

Brookes *et al.* (2004) discussed the complexity of pathogens being transported into lakes and reservoirs by rivers, considering that pathogens could settle to the bottom, or be inactivated by heat, ultraviolet light, and grazing by predators. Indicator bacteria could easily differ from pathogenic microbes in each of those ways. Skraber *et al.* (2004) compared coliform bacteria and coliphage viruses as indicators of viral contamination in river water. They found that 'the number of virus genome-positive samples decreased with decreasing concentration of coliphages, while no such relation was observed for thermotolerant coliforms'. The relationship between coliforms and pathogenic viruses varied with water temperature.

It would be interesting to know the differences between pathogens and indicators in their tendencies to be carried by water, their sedimentation rates, and their survival in waters of different types. Indicator bacteria are thought to die off faster in bright sun (Wymer *et al.*, 2004). For example, literature values indicate die-off rates for total coliform in darkness of 0–0.1 hr^{-1} (Hydroscience 1977) and in peak sunlight of 1.5–4.6 hr^{-1} (Fujioka *et al.*, 1981). A study by Noble and Fuhrman

(1997) found similar behavior for viruses in Santa Monica Bay, California. That study also found that particles and dissolved substances in seawater also affected virus decay; bacteria were apparently not studied simultaneously, however.

Sinton *et al.* (2002) studied differential inactivation by sunlight of fecal coliforms, enterococci, *E. coli*, somatic coliphages, and FRNA phages in fresh and marine waters. They found the somatic coliphages to be most sunlight-resistant in seawater and the FRNA phages most resistant in freshwater, and concluded that these would best represent enteric virus survival. Lemarchand *et al.* (2004) suggested that the bacterial indicators have greater sensitivity to disinfection than do viruses and protozoa cysts. However, Soller *et al.* (2003) reported that FRNA phages survived approximately four times longer in freshwater than did attenuated poliovirus.

3.5   Empirical relationships between indicators and pathogens

Of the studies measuring both indicator bacteria and pathogenic organisms in the same water samples, the majority have found little relationship. We describe some positive studies first, then studies that found no clear relationships.

3.5.1   Studies identifying positive relationships

Payment *et al.* (2000) studied pathogens and indicator levels at numerous sites on the St. Lawrence River. They found statistically significant rank correlations between three types of indicator bacteria (*Clostridium perfringens*, fecal coliforms, and total coliforms) and the pathogens *Cryptosporidium*, *Giardia*, and enteric viruses. They also found statistically significant relationships for the probability of occurrence of the three pathogen types when those were regressed logistically against the logarithms of the indicator concentrations. However, they stated that 'There was no lower level for any of the bacterial indicators at which the probability of finding a pathogen was zero'. Hörman *et al.* (2004) found no statistically significant correlations in the concentrations of thermotolerant coliforms, *E.coli*, or FRNA phages as indicators with the pathogens *Campylobacter* spp., *Giardia* spp., *Cryptosporidium* spp., and noroviruses in surface waters of Finland. They did find relationships between presence or absence of the indicators and presence or absence of the pathogens, but concluded that fecal indicator organisms alone were not sufficient for the assessment of the occurrence of a particular enteropathogen.

3.5.2   Studies not identifying positive relationships

Rose *et al.* (1987) found low correlations of 0.176 between enteroviruses and fecal coliforms and 0.060 between those viruses and fecal streptococcus in 18 water samples from an Arizona stream. That same study found enteroviruses and rotaviruses (sometimes at levels that exceeded state virus standards) in many samples that had levels of indicator bacteria considered acceptable. Kueh *et al.* (1995) found relationships between three genera of pathogenic bacteria and gastrointestinal symptoms in swimmers, but none between indicator bacteria and those symptoms. They did not analyze relationships between indicators and pathogens in their samples. Jiang *et al.* (2001) found that an 'excess of bacterial indicators did not correlate with the presence of human adenoviruses in coastal waters'.

Jiang and Chu (2004) used PCR to detect viruses in 21 samples from urban rivers in southern California, and measured indicator bacteria and coliphages as indicators in those samples. See Figure 3.3, in which each point represents one of the 21 water samples. Note that some of the *indicators* are roughly correlated with one another, especially enterococcus and fecal coliforms with total coliforms. Jiang and Chu concluded that 'There was no apparent relationship between the occurrence of human viruses and the microbial quality of the water based on indicators'. This point is generally supported by Figure 3.4, which shows the distributions of each indicator when each of three virus types was either present or absent. (In both these figures, the indicator concentrations are expressed on a $\log_{10}$ scale.)

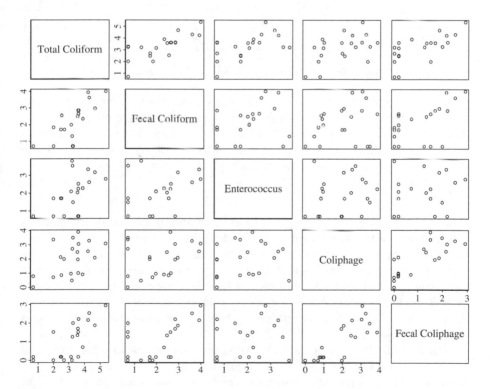

Figure 3.3   Pairs plot of data from Jiang and Chu (2004), for $\log_{10}$ concentrations of three indicator bacteria (total coliforms, fecal coliforms, and enterococci), and $\log_{10}$ concentrations of coliphages and fecal coliphages (as indicators).

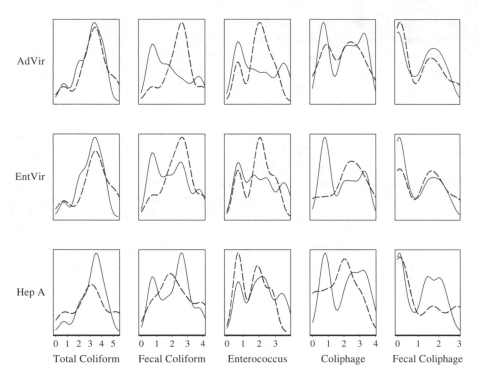

Figure 3.4   Empirical density curves showing the distributions of the five indicators ($\log_{10}$ concentrations), when each of three pathogenic virus groups was either detected (solid lines) or not detected (dashed lines), from data of Jiang and Chu (2004). AdVir = adenovirus, EntVir = enterovirus, and Hep A = hepatitis A. These curves may be thought of as 'smoothed histograms'.

Consider four examples. The top left-hand panel shows that the distributions of total coliforms were almost identical when adenovirus was detected as when it was not. Oddly, the next panel to the right shows that that indicator tended to have higher levels when adenovirus was absent than when it was present. On the other hand, concentrations of total coliforms and fecal coliphages had some tendency to be higher when hepatitis A was present than when it was not detected. Many of the other panels show no clear relationship between indicator levels and pathogen presence or absence.

Carter *et al.* (1987) sampled for the pathogen *Campylobacter jejuni* in natural waters in central Washington State and found no substantial or statistically significant correlations with total coliforms, fecal coliforms, fecal streptococci, or heterotrophic plate count. Griffin *et al.* (2003) noted that viable pathogenic human viruses can often be found in marine water influenced by human sewage. They went on to say that 'a common trend among many of the cited occurrence studies is that bacterial indicator occurrence did not correlate with viral occurrence', and that 'in a majority of the studies that monitored marine waters for both bacterial indicators and pathogenic viruses, viruses were detected when indicator levels were below public health water quality threshold levels'.

Lemarchand and Lebaron (2003) concluded that 'fecal indicators do not adequately indicate the presence of *Cryptosporidium* and *Salmonella* in natural waters and that pathogens and indicators may have different behaviors in the aquatic environment'. Arvanitidou *et al.* (1997) found that 'Mean log values of the standard indicator bacteria did not significantly differ between listeria and salmonella positive and negative samples'. Leclerc (2001) stated that 'It has been demonstrated beyond a doubt that, in spite of the undisputed merits of the different members of the coliform group,. . . they are marred by the essential shortcoming that their fate in the cycle of raw water to drinking water does not at all reflect adequate elimination of nonbacterial, waterborne pathogens'.

3.6 Non-fecal sources of indicators

Soil can be a major non-point source of fecal indicator bacteria in tropical surface waters (Hardina and Fujioka, 1991; Fujioka *et al.*, 1999; Solo-Gabriele *et al.*, 2000; Byappanahalli *et al.*, 2003; Whitman and Nevers, 2003), and studies in Hawaii, Guam, Puerto Rico, Indiana, south Florida, and Australia have suggested that *E. coli* can occur, and may even grow, in water, soils, sediments, beach sand, and plants (Carrillo *et al.*, 1985; Bermudez and Hazen, 1988; Hardina and Fujioka, 1991; Ashbolt *et al.*, 1997; Byappanahalli and Fujioka, 1998, 2004; Fujioka *et al.*, 1999; Solo-Gabriele *et al.*, 2000; Whitman and Nevers, 2003; Power *et al.*, 2005). In temperate habitats, *E. coli* has been found in wet sediments along streams (Whitman *et al.*, 1995, 1999; Byappanahalli *et al.*, 2003), nearshore beach sand (Alm *et al.*, 2003; Kinzelman *et al.*, 2003; Whitman and Nevers, 2003; Whitman *et al.*, 2003a), backshore, deep subsurface sand (Whitman *et al.*, 2003b), and forest soils (Byappanahalli *et al.*, 2003). Sediment-borne bacteria may be a source of increased levels of indicator bacteria found in surface waters (Davies *et al.*, 1995; An *et al.*, 2002).

Byappanahalli *et al.* (2006) recently monitored undisturbed, Dunes Creek forest soils in Indiana near Lake Michigan; *E. coli* was found in 88 % of the samples collected. The soil strains were genetically distinct from animal-related *E. coli* strains tested (i.e. gulls, terns, deer, and most geese). Although the primary source of *E. coli* is ill-defined, *E. coli* occurrence in these soils was widespread, persistent, and independent of short-term input. Once introduced, *E. coli* may remain viable for an extended period and can readily be transported to streams and adjacent waters by run-off, soil or sediment erosion, and aeolian processes. The idea that *E. coli* can persist and grow in soil is contrary to the generally held paradigm that fecal indicator bacteria are resident populations only in the intestinal tracts of warm-blooded animals (Byappanahalli *et al.*, 2006). These findings suggest that indicator bacteria found in surface waters may be unrelated to point-source sewage contamination, and when assessing the importance of *E. coli* found in surface waters and possible sources of contamination, especially for recreational waters, it is important to consider not only animal and human fecal contamination but also the contribution from background soil sources.

3.7   Epidemiologic studies of water quality indicators and recreational water health risks

Epidemiologic studies of adverse health effects associated with indicators of recreational water quality began in the 1950s when Stevenson (1953) studied three pairs of bathing sites on Lake Michigan, along the Ohio River, and on Long Island Sound. Eye, ear, nose, and throat symptoms, as well as gastrointestinal illness, were studied. Overall, there was no consistent correlation between illness and levels of coliform bacteria in the water; however, the data from Lake Michigan beaches suggested that increased illness was associated with increased levels of total coliforms. Prospective epidemiologic studies in the 1970s (Cabelli *et al.*, 1982) compared rates of gastrointestinal illness in swimmers and beach-going non-swimmers at beaches differing in microbial water quality and sources of fecal contamination. They concluded that levels of enterococci best correlated with gastrointestinal illness (e.g., vomiting, diarrhea, nausea, stomach ache) attributable to swimming in marine waters and that both enterococci and *E. coli* correlated with such illness in fecal contaminated freshwaters. Since then, other epidemiologic studies have been conducted throughout the world (NRC, 2004). The NRC (2004), WHO (2001), Prüss (1998), and Wade *et al.* (2003) have evaluated these recreational water studies. Some studies addressed a broad range of swimming-associated health including respiratory illness, some estimated the microbial quality of water to which bathers were exposed, and some measured microbial indicators such as coliphages, or pathogens such as enteric viruses and parasites. The WHO (2001) concluded that fecal streptococci and enterococci were the fecal indicator microorganisms that best predicted the relationships between swimming-associated health effects and the microbial quality of recreational waters.

Of the 37 studies reviewed by Prüss, 22 qualified for an evaluation by the National Research Council (NRC, 2004), which concluded that the rate of certain symptoms or groups of symptoms was significantly related to the count of fecal indicator bacteria in recreational water. There was a consistency across the various studies, with gastrointestinal symptoms being the most frequent health outcome for which significant dose-related associations were reported. In marine waters, fecal streptococci or enterococci were the fecal indicators that best predicted gastrointestinal illness. In freshwaters, increased concentrations of fecal coliforms or *E. coli* as well as fecal streptococci and enterococci were predictive of increased gastrointestinal illness risks. *Staphylococcus* concentrations were also found to be predictive of increased risks of illness, including ear, skin, respiratory, and gastrointestinal illness. The NRC noted that the relationships (temporal and dose–response) between microbial indicators and adverse health effects were strong and consistent and that the studies had biological plausibility and analogy to clinical cases from drinking contaminated water.

Wade *et al.* (2003) conducted a similar synthesis of 17 marine water and 10 freshwater studies. The review included 24 cohort studies, two randomized trials, and one case–control study. A subset of the studies was amenable to the determination of the relative risk of a health outcome, such as gastrointestinal, respiratory, skin, ear, or eye effects. Levels of the following water quality indicators in recreational water were significantly associated with gastrointestinal illness at one or more beaches: fecal streptococci, enterococci, fecal coliforms, *E. coli*, total coliforms (marine water only), enteroviruses (marine water only), and coliphages (freshwater only). *E. coli* and enterococci were the best (i.e., most consistent) indicators of gastrointestinal illness in marine water; there was no consistent indicator of gastrointestinal illness in freshwater. The investigators also noted that enteroviruses and bacteriophages may be promising indicators to predict risk of gastrointestinal illness, but there were too few studies to establish their utility. None of the commonly used microbial indicators consistently predicted risks of respiratory illness, but relative risks of skin disorders tended to increase with several indicators, including fecal streptococci, enterococci, fecal coliforms, and *E. coli*.

The results of the epidemiologic studies provide encouraging evidence that predictive associations exist or can be found between various swimming-associated health effects and microbial indicators in recreational bathing waters. However, several research gaps were identified (NRC, 2004) regarding the use of indicators for recreational waters, including the following gaps:

- studies of immunocompromised populations and other sensitive/vulnerable subpopulations such as children and the elderly;
- studies using enteric viruses and bacteriophages as water quality indicators and combinations of water quality indicators to assess overall health risks; and
- analyses of the effects of study location and climate on results.

In addition, other candidate water quality indicators have not been adequately studied, or reliable methods were previously not available to include these indicators in the epidemiologic studies.

3.8   Conclusions

Although it is reasonable to suppose that the bacteria commonly used to indicate the presence of fecal material in water would be closely related to pathogenic organisms released in the feces of humans and other animals, the fates and transport of indicators and pathogens apparently differ substantially in the environment. The majority of studies we reviewed found little quantitative relationship between those two types of microbes, although a few did find presence–absence relationships. In addition, a commonly used indicator, *E. coli*, may persist and grow in soil and be carried into surface waters during rain events.

Despite a common lack of correlation between indictor and pathogen densities, reviews of epidemiological studies by Prüss (1998), Wade *et al.* (2003), and the NRC (2004) have concluded that low levels of indicator bacteria in recreational waters are sufficiently predictive of low illness rates to be useful in helping to protect the health of swimmers. These reviews also found considerable variation in the estimated *quantitative* relationships between indicator organism densities and rates of adverse health effects for swimmers. Given the many potential sources of variability in the relationships between indicators and actual pathogen health risks, some of which we have described, the substantial variation in the quantitative relations between indicator densities and rates of adverse health effects is not surprising. A major question facing regulators in using epidemiological data obtained at beaches is whether the indicator–illness relationships found at studied beaches have enough in common among themselves, and with beaches not yet studied, to allow setting criteria for safety that will be sufficiently protective in general, without being overly protective at some locations.

Attempts to derive widely applicable exposure–risk curves, e.g., one for fresh-waters and one for marine waters, in which the risk to public health is deemed acceptable below a single value, seem inappropriate given the substantial variability and uncertainty that is known to be present. The great variation of the exposure–risk curves in Figures 3 and 4 of Prüss (1998), in which indicator concentrations associated with specific degrees of risk vary over several orders of magnitude in different studies, suggests that the exposure–risk relationship will vary not only from time to time but also from beach to beach If this is the case, then a single guideline or regulatory value may not apply well to all beaches. Rather than using a threshold level of indicator organism density as an acceptable limit based on a linear or log-linear indicator–illness function, it would seem more realistic to think in terms of exposure–risk *bands,* which would account for the variability and uncertainty of indicator–risk relationships across time and among beaches. Such bands might need to be quite wide.

Put another way, if regulators were to adopt a flagging system to indicate beach safety—green for fairly certain low risk, yellow for intermediate and uncertain risk, and red for fairly certain high risk—then yellow flags would likely be common. Given the variability known to occur among beaches and through time, this type of approach to setting criteria seems worth considering.

Finally, we note that some authors (e.g., Havelaar *et al.*, 1993; Sinton *et al.*, 2002; Cole *et al.*, 2003; Skraber *et al.* 2004) have suggested using phages rather than bacteria as indicators, as the viral phages appear to be more closely related to viral pathogens than bacteria are. For the future, Lemarchand *et al.* (2004) suggested that DNA microarray technology allows detection of numerous DNA sequences simultaneously, and might be used to detect the presence of multiple pathogens from single samples. Given the fast-developing power of molecular biological methods, further research on methods for detecting suites of actual pathogens in recreational waters seems worthwhile. In any case, epidemiologic studies would be required to demonstrate that such new indicators, even if they are more closely related to pathogen levels than those used to date, would predict illness risk effectively.

Acknowledgments

We thank Alexandra Boehm and Larry Wymer for helpful suggestions that have improved this chapter.

References

Adam, R.D. (2000) The *Giardia lamblia* genome. *International Journal for Parasitology*, 30: 475–484.

Agarwal, R.K., Kapoor, K.N., and Kumar, A. (1998) Virulence factors of aeromonads – an emerging food borne pathogen problem. *Journal of Communicable Diseases*, 30(2): 71–78.

Allen, M.J., and Geldreich, E.E. (1978) Evaluating the microbial quality of potable waters. In *Evaluation of the Microbiology Standards for Drinking Water* (EPA 570/9-78-OOC). EPA, Washington, DC, pp. 37–48.

Allwood, P.B., Malik, Y.S., Hedberg, C.W., and Goyal, S.M. (2003) Survival of F-specific RNA coliphage, feline calicivirus, and *Escherichia coli* in water: a comparative study. *Applied and Environmental Microbiology*, 69(9): 5707–5710.

Alm, E.W., Burke, J., and Spain, A. (2003) Fecal indicator bacteria are abundant in wet sand at freshwater beaches. *Water Research*, 37: 3978–3982.

An, Y.-J., Kampbell, D.H., and Breidenbach, G.P. (2002) *Escherichia coli* and total coliforms in water and sediments at lake marinas. *Environmental Pollution*, 120: 771–778.

Arvanitidou, M., Papa, A., *et al.* (1997). The occurence of *Listeria* spp. and *Salmonella* spp. in surface waters. *Microbiological Research*, 152: 395–397.

Ashbolt, N.J., Dorsch, M.R., Cox, P.T., and Banes, B. (1997) Blooming of *E. coli*, what do they mean? In D. Kay and C. Fricker (eds), *Coliforms and E. Coli, Problem or Solution?* Royal Society of Chemistry, Cambridge.

Batik, O., Craun, G.F., and Pipes, W.O. (1983) Routine coliform monitoring and waterborne disease outbreaks. *Journal of Environmental Health*, 45: 227.

Benenson, A.S. (1995) *Control of Communicable Disease Manual*, 16th edition. American Public Health Association, Washington, DC.

Bennett, J.V., Holmberg, S.D., Rogers, M.F., and Solomon, S.L. (1987) Infectious and parasitic diseases. In R.W. Amler and H.B. Dull (eds), *Closing the Gap: The Burden of Unnecessary Illness*. Oxford University Press, New York.

Bermudez, M., and Hazen, T.C. (1988) Phenotypic and genotypic comparison of *Escherichia coli* from pristine tropical waters. *Applied and Environmental Microbiology*, 54: 979–983.

Bottone, E.J. (1999) *Yersinia enterocolitica*: overview and epidemiologic correlates. *Microbes and Infection*, 1(4): 323–333.

Brookes, J.D., Antenucci, J., Hipsey, M., Burch, M.D., Ashbolt, N.J., and Ferguson, C. (2004) Fate and transport of pathogens in lakes and reservoirs. *Environment International*, 30: 741–759.

Byappanahalli, M.N., and Fujioka, R.S. (1998) Evidence that tropical soil environment can support the growth of *Escherichia coli*. *Water Science and Technology*, 38: 171–174.

Byappanahalli, M.N., and Fujioka, R.S. (2004) Indigenous soil bacteria and low moisture may limit but allow faecal bacteria to multiply and become a minor population in tropical soils. *Water Science and Technology*, 50: 27–32.

Byappanahalli, M., Fowler, M., Shively, D., and Whitman, R. (2003) Ubiquity and persistence of *Escherichia coli* in a midwestern coastal stream. *Applied and Environmental Microbiology*, 69: 4549–4555.

Byappanahalli, M.N., Whitman, R.L., Shively, D.A., Sadowsky, M.J., and Ishii, S. (2006) Population structure, persistence, and seasonality of autochthonous Escherichia coli in temperate, coastal forest soil from a Great Lakes watershed. *Environmental Microbiology*, 8(3): 504–513.

Cabelli, V.J. (1983) Health Effects Criteria for Marine Recreational Waters. Cincinnati, OH: EPA-600-1-80-031.

Cabelli, V.J., Dufour, A.P., McCabe, L.J., and Levin, M.A. (1982) Swimming-associated gastroenteritis and water quality.*American Journal of Epidemiology* 115: 606–616.

Carrillo, M., Estrada, E., and Hazen, T.C. (1985) Survival and enumeration of the fecal indicators *Bifidobacterium adolescentis* and *Escherichia coli* in a tropical rain forest watershed. *Applied and Environmental Microbiology*, 50: 468–476.

Carter, A.M., Pacha, R.E., Clark, G.W., and Williams, E.A. (1987) Seasonal occurence of *Campylobacter* spp. in surface waters and their correlation with standard indicator bacteria. *Applied and Environmental Microbiology*, 53: 523–526.

Cole, D., Long, S.C., and Sobsey, M.D. (2003) Evaluation of F+ RNA and DNA coliphages as source-specific indicators of fecal contamination in surface waters. *Applied and Environmental Microbiology*, 69(11): 6507–6514.

Cotruvo, J.A., Dufour, A., Rees, G., Bartrum, J., Carr, R., Cliver, D.O., Craun, G.F., Fayer, R., and Gannon, V.P.J. (2004) *Waterborne Zoonoses: Identification, Causes and Control.* IWA Publishing, London.

Craun, G.F., Berger, P.S., Calderon, R. (1997) Coliform bacteria and waterborne disease outbreaks. *Journal of the American Water Works Association*, 89(3): 96–104.

Craun, G.F., Calderon, R., Craun, M.F. (2004) Waterborne outbreaks caused by zoonotic pathogens in the USA. In J.A. Cotruvo, A. Dufour, G. Rees, J. Bartrum, R. Carr, D.O. Cliver, G.F. Craun, R. Fayer, and V.P.J. Gannon (eds), *Waterborne Zoonoses: Identification, Causes and Control.* IWA Publishing, London.

Craun, G.F., Calderon, R., and Craun, M.F. (2005) Outbreaks associated with recreational water in the United States. *International Journal of Environmental Health Research*, 15(4): 243–262.

Davies, C.M., Long, J.A.H., Donald, M., and Ashbolt, N.J. (1995) Survival of fecal microorganisms in marine and freshwater sediments. *Applied and Environmental Microbiology* 61: 1888–1896.

Dufour, A. (1984) Health effects criteria for fresh recreational waters. US Environmental Protection Agency (EPA-600/1-84-004), Research Triangle Park, NC.

Dupont, H., Chappell, C., Sterling, C., Okhuysen, P., Rose, J., and Jakubowski, W. (1995) The infectivity of *Cryptosporidium parvum* in healthy volunteers. *New England Journal of Medicine*, 332: 855.

Fankhauser, R.L., Noel, J.S., Monroe, S.S., Ando, T., and Glass, R.I. (1998) Molecular epidemiology of 'Norwalk-like viruses' in outbreaks of gastroenteritis in the United States. *Journal of Infectious Diseases* 178(6): 1571–1578.

Feachem, R.G., Bradley, D.J., Garelick, H., and Mara, D.D. (1983) *Sanitation and Disease: Health Aspects of Excreta and Wastewater Management.* John Wiley & Sons, Ltd., Chichester.

Frost, F.J., Muller, T., Kunde, T., Craun, G.F., and Calderon, R.L. (2003) Seroepidemiology. In P.R. Hunter, M. Waite, and E. Ronchi (eds), *Drinking Water and Infectious Disease: Establishing the Links.* CRC Press, Boca Raton, FL.

Frost, F., Roberts, M., Kunde, T., Craun, G., Tollestrup, K., Harter, L. and Muller, T. (2005) How clean must our drinking water be: the importance of protective immunity. *Journal of Infectious Diseases*, 191: 809–814.

Fujioka, R.S., Hashimoto, H.H., Siwak, E.B., and Young, R.H. (1981) Effect of sunlight on survival of indicator bacterial in seawater. *Applied and Environmental Microbiology*, 41(3): 690–696.

Fujioka, R., Sian-Denton, C., Borja, M., Castro, J., and Morphew, K. (1999) Soil: the environmental source of *Escherichia coli* and enterococci in Guam's streams. *Journal of Applied Microbiology* (Symposium Supplement), 85: 83S–89S.

Geldreich, E.E. (1991) Microbial water quality concerns for water supply use. *Environmental Toxicology and Water Quality*, 6: 209–223.

Griffin, D.W., Donaldson, K.A., Paul, J.H., and Joan B. Rose, J.B. (2003) Pathogenic human viruses in coastal waters. *Clinical Microbiology Reviews*, 16(1): 129–143.

Haas, C.N., Rose, J.B., and Gerba, C.P. (1999). *Quantitative Microbial Risk Assessment.* John Wiley & Sons, Inc., New York.

Hardina, C.M., and Fujioka, R.S. (1991) Soil: the environmental source of *Escherichia coli* and enterococci in Hawaii's streams. *Environmental Toxicology and Water Quality*, 6: 185–195.

Havelaar, A.H., Olphen, M.V., and Drost, Y.C. (1993) F-specific RNA bacteriophages are adequate model organisms for enteric viruses in fresh water. *Applied and Environmental Microbiology*, 95: 2956–2962.

Hoebe, C.J.P.A., Vennema, H., Husman, A.M.D. and van Duynhoven, Y.T.H.P. (2004) Norovirus outbreak among primary schoolchildren who had played in a recreational water fountain. *Journal of Infectious Diseases*, 189(4): 699–705.

Hörman, A., Rimhanen-Finne, R., Maunula, L., von Bonsdorff, C.-H., Torvela, N., Heikinheimo, A., and Marja-Liisa Hänninen, M.-L. (2004) *Campylobacter* spp., *Giardia* spp., *Cryptosporidium* spp., noroviruses, and indicator organisms in surface water in southwestern Finland, 2000–2001. *Applied and Environmental Microbiology*, 70(1): 87–95.

Hutson, A.M., Atmar, R.L., Graham, D.Y., and Estes, M.K. (2002) Norwalk virus infection and disease is associated with ABO histo-blood group type. *Journal of Infectious Diseases*, 185(9): 1335–7.

Hydroscience (1977) Task report, special water quality studies (PCP Task 317). City of New York, 208 Report, Hazen & Sawyer, Managing Consultants, New York.

Isonhood, J.H., and Drake, M. (2002) *Aeromonas* species in foods. *Journal of Food Protection*, 65(3): 575–582.

Jiang, S.C. and Chu, W. (2004). PCR detection of pathogenic viruses in southern California urban rivers. *Journal of Applied Microbiology*, 97: 17–28.

Jiang, S., Noble, R., and Chu, W. (2001). Human adenoviruses and coliphages in urban runoff-impacted coastal waters of southern California. *Applied and Environmental Microbiology*, 67(1): 179–184.

Kinzelman, J., Whitman, R.L., Byappanahalli, M., Jackson, E., and Bagley, R.C. (2003) Evaluation of beach grooming techniques on *Escherichia coli* density in foreshore sands at North Beach, Racine, WI. *Lake and Reservoir Management*, 19: 349–354.

Kueh, C.S.W., Tam, T.-Y., Lee, T., Wong, S.L., Lloyd, O.L., Yu, I.T.S., Wong, T.W., Tam, J.S., and Bassett, D.C.J. (1995). Epidemiological study of swimming-associated illnesses relating to bathing-beach water quality. *Water Science and Technology*, 31(5–6): 1–4.

Leclerc, H., Mossel, D.A., Edberg, S.C., and Struijk, C.B. (2001). Advances in the bacteriology of the coliform group: their suitability as markers of microbial water safety. *Annual Review of Microbiology*, 55: 201–234.

Lemarchand, K., and Lebaron, P. (2003) Occurrence of *Salmonella* spp. and *Cryptosporidium* spp. in a French coastal watershed: relationship with fecal indicators. *FEMS Microbiology Letters*, 218: 203–209.

Lemarchand, K., Masson, L., and Brousseau, R. (2004) Molecular biology and DNA microarray technology for microbial quality monitoring of water. *Critical Reviews in Microbiology*, 30: 145–172.

Levine, W.C., and Stephenson, W.T. (1990) Waterborne disease outbreaks, 1986–1988. *Morbidity and Mortality Weekly Report*, 39(SS-1).

Lindesmith, L., Moe, C., Marionneau, S., Ruvoen, N., Jiang, X., Lindbland, L., Stewart, P., LePendu, J., and Baric, R. (2003) Human susceptibility and resistance to Norwalk virus infection. *Nature Medicine*, 9(5): 548–553.

Lodder, W.J., and Husman, A.M. de R. (2005) Presence of noroviruses and other enteric viruses in sewage and surface waters in The Netherlands. *Applied and Environmental Microbiology*, 71(3): 1453–1461.

Lopman, B.A., Brown, D.W., and Koopmans, M. (2002) Human caliciviruses in Europe. *Journal of Clinical Virology*, 24 (3): 137–160.

Marionneau, S., Ruvoen, N., Le Moullac-Vaidye, B., Clement, M., Cailleau-Thomas, A., Ruiz-Palacois, G., Huang, P., Jiang, X., and Le Pendu, J. 2002. Norwalk virus binds to histo-blood group antigens present on gastroduodenal epithelial cells of secretor individuals. *Gastroenterology*, 122(7): 1967–1977.

Maunula, L., Kalso, S., von Bonsdorff, C.H., and Ponka, A. (2004) Wading pool water contaminated with both noroviruses and astroviruses as the source of a gastroenteritis outbreak. *Epidemiology and infection*, 132(4): 737–743.

McBride, G., Till, D., Ryan, T., Ball, A., Lewis, G., Palmer, S., and Weinstein, P. (2002) Pathogen occurrence and human health risk assessment analysis. Freshwater Microbiology Research Programme Report, New Zealand Ministry for the Environment.

McCabe, L.J. (1977) Epidemiological consideration in the application of indicator bacteria in North America. In A.W. Hoadley and B.J. Dutka (eds), *Bacterial Indicators/Health Hazards Associated with Water, ATSM STP 635*. American Society for Testing and Materials, Philadelphia, pp. 15–22.

McCullough, N.B. and Eisele, C.W. (1951) Experimental human salmonellosis. I. Pathogenicity of strains of *Salmonella meleagridis* and *Salmonella anatum* obtained from spray-dried whole egg. *Journal of Infectious Diseases*, 88(3): 278–289.

McFeters, G.A., Schillinger, J.E., and Stuart, D.G. (1978) Alternative indicators of water contamination and some physiological characteristics of heterotrophic bacteria in water. In *Evaluation of the Microbiology Standards for Drinking Water* (EPA 570/9-78-OOC). EPA, Washington, DC, pp. 3–11.

Mead, P.S., Slutsker, L., Dietz, V., McCaig, L.F., Bresee, J.S., Shapiro, C., Griffin, P.M., and Tauxe, R.V. (1999) Food related illness and death in the United States. Emerging Infectious Diseases, 5(5): 607–625.

Morgan, U.M., Monis, P.T., Fayer, R., and Deplazes, P. (1999) Phylogenetic relationships among isolates of *Cryptosporidium*: evidence for several new species. *Journal of Parasitology*, 85(6): 1126–1133.

National Research Council (1977) *Drinking Water and Health.* National Academy of Sciences, Washington, DC.

National Research Council (2004) *Indicators for Waterborne Pathogens.* National Academies Press, Washington, DC, pp. 80–89.

Noble, R.T., and Fuhrman, J.A. (1997) Virus decay and its causes in coastal waters. *Applied and Environmental Microbiology,* 63(1): 77–83.

Nwachuku, N., Craun, G.F., and Calderon, R.L. (2002) How effective is the TCR in assessing outbreak vulnerability. *Journal of the American* Water Works Association, 94(9): 88–96.

Okhuysen, P.C. and Chappell, C.L. (2002) *Cryptosporidium* virulence determinants – are we there yet? *International Journal for Parasitology,* 32(5): 517–525.

Okhuysen, P.C., Chappell, J.H., Crabb, J.H., Sterling, C.R., and DuPont, H.L. (1999) Virulence of three distinct *Cryptosporidium parvum* isolates for healthy adults. *Journal of Infectious Diseases,* 180(4): 1275–1281.

Okhuysen, P.C., Rich, S.M., Chappell, C.L., Grimes, K.A., Widmer, G., Feng, X., and Tzipori, S. (2002) Infectivity of a *Cryptosporidium parvum* isolate of cervine origin for healthy adults and interferon-gamma knockout mice. *Journal of Infectious Diseases,* 185(9): 1320–1325.

Olivieri A.W. and Soller, J.A. (2001) Evaluation of the public health risks concerning infectious disease agents associated with exposure to treated wastewater discharged by the City of Vacaville, Easterly Wastewater Treatment Plant. Prepared by EOA Inc. for the City of Vacaville.

Palmateer, G.A., Dutka, B.J., Janzen, E.M., Meissner, S.M., and Sakellaris, M.G. (1991) Coliphage and bacteriophage as indicators of recreational water quality. *Water Research,* 25(3), 355–357.

Payment, P., Berte, A., Prevost, M., Menard, B., and Barbeau, B. (2000) Occurrence of pathogenic microorganisms in the Saint Lawrence River (Canada) and comparison of health risks for populations using it as their source of drinking water. *Canadian Journal of Microbiology,* 46: 565–576.

Power, M.L., Littlefield-Wyer, J., Gordon, D.M., Veal, D.A., and Slade, M.B. (2005) Phenotypic and genotypic characterization of encapsulated *Escherichia coli* isolated from blooms in two Australian lakes. *Environmental Microbiology,* 7: 631–640.

Prüss, A. (1998) Review of epidemiological studies on health effects from exposure to recreational water. *International Journal of Epidemiology,* 27: 1–9.

Rose, J.B., Mullinax, R.L., Singh, S.N., Yates, M.V. and Gerba, C.P. (1987) Occurrence of rotaviruses and enteroviruses in recreational waters of Oak Creek, Arizona. *Water Research,* 21(11): 1375–1381.

Ryan, J.K. (ed.) (1994) *Sherris Medical Microbiology: An Introduction to Infectious Diseases,* 3rd edition. Appleton & Lange, Norwalk, CT.

Schaechter, M. (1998) *Mechanisms of Microbial Disease,* 3rd edition. Williams & Wilkins, Baltimore, MD.

Schiff, G.M., Stefanovic, G.M., Young, E.C., Sander, D.S., Pennekamp, J.K. and Ward, R.L. (1984) Studies of echovirus-12 in volunteers: determination of minimal infectious dose and the effect of previous infection on infectious dose. *Journal of Infectious Diseases,* 150: 858–866.

Sinton, L.W., Hall, C.H., Lynch, P.A., and Davies-Colley, R.J. (2002) Sunlight inactivation of faecal indicator bacteria and bacteriophages from waste stabilization pond effluent in fresh and saline waters. *Applied and Environmental Microbiology,* 68(3): 1122–1131.

Skraber, S., Gassilloud, B., and Gantzer, C. (2004). Comparison of coliforms and coliphages as tools for assessment of viral contamination in river water. *Applied and Environmental Microbiology,* 70(6): 3644–3649.

Sobsey, M., Battigelli, D., Handzel, T., and Schwab, K. (1995) Male-specific coliphages as indicators of viral contamination of drinking water. *American Water Works Association Research Foundation, Denver, CO.*

Soller, J.A., Olivieri, A., Crook, J., Cooper, R.C., Tchobanoglous, G., Parkin, R.T., Spear, R.C. and Eisenberg, J.N.S. (2003). Risk-based approach to evaluate the public health benefit of additional wastewater treatment. *Environmental Science and Technology,* 37(9): 1882–1891.

Soller, J.A., Olivieri, A.W., Eisenberg, J.N.S., Sakaji, R., and Danielson, R. (2004) Evaluation of microbial risk assessment techniques and applications. *Water Environment Research Foundation Report,* Project 00-PUM-3.

Solo-Gabriele, H.M., Wolfert, M.A., Desmarais, T.R., and Palmer, C.J. (2000) Sources of *Escherichia coli* in a coastal subtropical environment. *Applied and Environmental Microbiology,* 66: 230–237.

Stephen, J. (2001) Pathogenesis of infectious diarrhea. *Canadian Journal of Gastroenteology,* 15(10): 669–83.

Stevenson, A.H. (1953) Studies of bathing water quality and health. *American Journal of Public Health,* 43 (5): 529–538.

Sulaiman, I.M., Xiao, L., and Lal, A. (1999) Evaluation of *C. parvum* genotyping techniques. *Applied and Environmental Microbiology,* 65(10): 4431–4432.

Sulaiman, I.M., Morgan, U.M., Thompson, R.C.A., Lal, A.A., and Xiao, L. (2000) Phylogenetic relationships of *Cryptosporidium* parasites based on the 70-kilodalton heat shock protein (HSP70) gene. *Applied and Environmental Microbiology,* 66: 2385–2391.

Toranzos, G.A., and McFeters, G.A. (1997) Detection of indicator microorganisms in environmental freshwaters and drinking waters. In C. Hurst, G. Knudsen, M. McInerney, L. Stetzenbach, and M.V. Walter (eds), *Manual of Environmental Microbiology.* American Society for Microbiology Press, Washington, DC.

US Environmental Protection Agency (1975) 40 CFR Part 141. Water programs: national interim primary drinking water regulations. *Federal Register,* 40: 59566–59574.

US Environmental Protection Agency (1986) Ambient water quality criteria for bacteria (EPA 440/5-84-002).

US Environmental Protection Agency (1989) 40 CFR Parts 141 and 142. Drinking water programs: national primary drinking water regulations; total coliforms (including fecal coliforms and *E. coli)*; final rule. *Federal Register,* 54: 27544–27568.

US Environmental Protection Agency (1990) 40 CFR Parts 141 and 142. Drinking water programs: national primary drinking water regulations; total coliforms; corrections and technical amendments; final rule. *Federal Register,* 55: 25064–25065.

US Environmental Protection Agency (2002a) Summary of the August 14–15, 2002, Experts Workshop on Public Health Impacts of Sewer Overflows (EPA833-R-02-02).

US Environmental Protection Agency (2002b) Implementation guidance for ambient water quality criteria for Bacteria (EPA-821-B-02-003).

Wade, T.J., Pai, N., Eisenberg, J.N.S., and Colford Jr., J.M. (2003) Do U.S. Environmental Protection Agency water quality guidelines for recreational waters prevent gastrointestinal illness? A systematic review and meta-analysis. *Environmental Health Perspectives,* 111(8): 1102–1109.

Ward, R.L., Bernstein, D.L., Young, E.C., Sherwood, J.R., Knowlton, D.R., and Schiff, G.M. (1986) Human rotavirus studies in volunteers: determination of infectious dose and serological response to infection. *Journal of Infectious Diseases,* 154(5): 871–880.

Whitman, R.L., and Nevers, M.B. (2003) Foreshore sand as a source of *Escherichia coli* in nearshore water of a Lake Michigan beach. *Applied and Environmental Microbiology*, 69: 5555–5562.

Whitman, R.L., Gochee, A.V., Dustman, W.A., and Kennedy, K.J. (1995) Use of coliform bacteria in assessing human sewage contamination. *Natural Areas Journal*, 15: 227–233.

Whitman, R.L., Nevers, M.B., and Gerovac, P.J. (1999) Interaction of ambient conditions and fecal coliform bacteria in southern Lake Michigan waters: monitoring program implications. *Natural Areas Journal*, 19: 166–171.

Whitman, R.L., Shively, D.A., Pawlik, H., Nevers, M.B., and Byappanahalli, M.N. (2003a) Occurrence of *Escherichia coli* and enterococci in *Cladophora* (Chlorophyta) in nearshore water and beach sand of Lake Michigan. *Applied and Environmental Microbiology*, 69: 4714–4719.

Whitman, R.L., Shively, D.A., Nevers, M.B., Korinek, G.C., and Byappanahalli, M.N. (2003b) Persistence of *E. coli* and enterococci in deep, backshore sand of two southern Lake Michigan beaches. *Abstracts of the International Water Association Symposium on Health-Related Water Microbiology*, Cape Town, South Africa, September 14–19.

Widmer, G., Tzipori, S., Fichtenbaum, C., and Griffiths, J.K. (1998) Genotypic and phenotypic characterization of *Cryptosporidium parvum* isolates from people with AIDS. *Journal of Infectious Diseases*, 178(7): 834–840.

World Health Organization (1999) Health based monitoring of recreational waters: the feasibility of a new approach (The 'Annapolis Protocol'), WHO/SDE/WSH/99.1. Geneva.

World Health Organization (2001) Bathing water quality and human health: faecal pollution. Outcome of an Expert Consultation, Farnham, UK, April. Cosponsored by Department of the Environment, Transport and the Regions, United Kingdom. WHO/SDE/WSH/01.2. WHO, Geneva.

Wymer, L.J., Dufour, A.P., Brenner, K.P., Martinson, J.W., Stutts, W.R., and Schaub, S.A. (2004) The EMPACT Beaches Project. Results from a Study on Microbiological Monitoring In Recreational Waters (US EPA 600/R 04/023). Cincinnati, OH.

Xiao, L., Morgan, U.M., Fayer, R., Thompson, R.C.A., and Lal, A.A. (2000) *Cryptosporidium* systematics and implications for public health. *Parasitology Today*, 16(7): 287–292.

Yates, M.V., and Gerba, C.P. (1998) Microbial considerations in wastewater reclamation and reuse. In T. Asano (ed.), *Wastewater Reclamation and Reuse*. Technomic Publishing, Lancaster, PA, 437–488.

Yoder, J.S., Blackburn, B.G., Craun, G.F., *et al.* (2004) Surveillance for waterborne-disease outbreaks associated with recreational water – United States, 2001–2002. *Morbidity and Mortality. Weekly Report*, 53(SS-8): 1–22.

4

On selecting the statistical rationale for revised EPA recreational water quality criteria for bacteria

Richard O. Gilbert

Environmental Statistician, Rockville, MD, USA

4.1 Introduction

It is generally assumed that swimmers at fresh and marine recreational beaches are susceptible to increased likelihood of gastrointestinal illness when fecal bacteria counts in the beach water reach certain levels. The quality of beach and other recreational waters must be monitored to determine whether it is safe to swim. The US Congress and the US Environmental Protection Agency (EPA) have taken actions to improve health and environmental protection programs for beachgoers and provide the public with information about the quality of their beach waters. In 1997 the EPA established the Beaches Environmental Assessment and Coastal Health (BEACH) program. In 1999 the EPA published the *Action Plan for Beaches and Recreational Waters* (US EPA, 1999), and in October 2000, the US Congress passed the BEACH Act, which established certain EPA BEACH Program activities as statutory requirements.

Statistical Framework for Recreational Water Quality Criteria and Monitoring   Edited by Larry J. Wymer
© 2007 John Wiley & Sons, Ltd

The BEACH Act requires the EPA to develop new, improved criteria for pathogens and pathogen indicators. The EPA will soon revise the ambient water quality criteria for bacteria in fresh and marine recreational waters in accordance with the Clean Water Act 1977 and the BEACH Act 2000. The criteria published by EPA guide the states in setting their own criteria, as required by Section 304 of the Clean Water Act.

This chapter considers whether statistical process control (PC) and *control charts* (CCs) or *acceptance sampling* (AC) should be the underlying philosophy or rationale adopted by the EPA for the new ambient water quality criteria for bacteria. The current EPA guidance (US EPA, 1986, p. 9) uses a *control chart* analogy in setting maximum density count limits for individual measurements of beach water samples. On the other hand, Wymer *et al.* (2004) discuss the use of AC as a possible framework for deciding if density limits have been exceeded. The main difference in philosophy is that CCs focus more on historical data collected over longer periods of time to make decisions than does AC.

We begin in Section 4.2 by reviewing the current EPA water quality criteria for bacteria published in US EPA (1986). Section 4.3 discusses some factors that the EPA may choose to consider when selecting the decision-making rationale. Section 4.4 reviews and discusses CCs, while Sections 4.5 and 4.6 discuss and illustrate AC plans. Section 4.7 provides discussion.

4.2   Current EPA Criteria

Section 304(a)(1) of the Clean Water Act 1977 requires that the EPA publish criteria for water quality. The current criteria (US EPA, 1986) are based on a series of studies at marine and freshwater swimming beaches conducted in the 1970s (Cabelli, 1983; Dufour, 1984). The criteria have two components: (1) the particular indicator organisms for bacteria that should be monitored in beach waters; and (2) the numerical limits for each indicator. US EPA (1986) recommended using the fecal coliform bacteria *Escherichia coli* or the intestinal bacteria enterococci (not both) for freshwater beaches, and enterococci for marine beaches.

The criteria published in US EPA (1986) are provided in Table 4.1. Maximum numerical limits are given for both the average density (geometric mean, GM) and individual colony-forming unit (CFU) measurements. Non-compliance with the criteria in Table 4.1 is in effect when the computed GM exceeds the numerical limit for the GM or when any individual density measurement exceeds the maximum acceptable density for individual measurements. Figure 4.1 shows three hypothetical data sets obtained for three different time periods. For time period 1 the data indicate the US EPA (1986) criteria have been met, whereas the beach is in non-compliance for periods 2 and 3. US EPA (1986, p. 16) stipulates that the GM is to be computed based on a statistically sufficient number of samples (generally not less than five samples equally spaced over a 30-day period).

The GM limits were computed by US EPA (1986, p. 6) using an estimated linear regression of swimming-associated gastrointestinal illness rates on GM indicator

Table 4.1 US EPA (1986) water quality criteria (colony-forming units per 100 ml) for bacteria at recreational beaches

Indicator organisms that measure the presence or absence of fecal bacteria in beach waters	Acceptable swimming-associated gastroenteritis rate per 1000 swimmers[a]	Steady-state geometric mean maximum acceptable indicator density	Single-measurement maximum acceptable density			
			Heavy-use swimming beach area (upper 75th percentile)[b]	Moderate-use swimming beach area (upper 82nd percentile)[b]	Light-use swimming beach area (upper 90th percentile)[b]	Infrequently used swimming beach area (upper 95th percentile)[b]
Fresh water						
Enterococci	8	33	61	78	107	151
E. coli	8	126	235	298	409	575
Marine water						
Enterococci	19	35	104	158	276	501

[a]The numerical water quality criteria in this table approximately correspond to these acceptable gastroenteritis rates based on linear regression equations developed by Cabelli (1983) and Dufour (1984).

[b]The single-sample maximum acceptable density was computed in US EPA (1986, equation 4.4 in Table 4) as the designated percentile of a lognormal distribution with an assumed standard deviation of the logarithms of the counts of 0.4 for freshwater beaches and 0.7 for marine beaches.

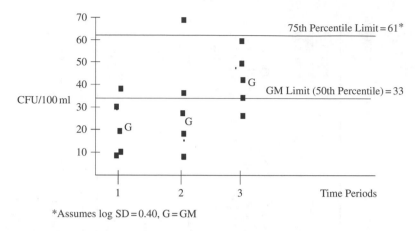

Figure 4.1   Three hypothetical data sets. The beach is not in compliance with US EPA (1986) criteria for the periods 2 and 3.

densities. The individual measurement limits were computed using the equation for computing percentiles of a lognormal distribution (US EPA, 1986, note 4 in Table 4, page 15). US EPA (1986) discusses these percentiles in general terms as 'confidence limits'. However, it should be understood that the percentiles are simply CFU values at the appropriate percentile locations on a lognormal distribution curve. They are not confidence limits in the usual statistical definition of that term.

The individual measurement limits correspond to the 75th, 82nd, 90th and 95th percentiles of the assumed lognormal distribution to which the GM limit applies. (Recall that a percentile of a population, say the 75th percentile, is the value below which 75 % of the individual measurements of the population lie and above which 25 % of the measurements lie.) The individual measurement limits were computed by the EPA assuming that the standard deviation (SD) of the logarithms of density counts obtained on representative water samples (the 'log SD') equals 0.40 and 0.70 for freshwater and marine water beaches, respectively. These computed log SDs were obtained for beach water density counts obtained over several days for the beaches studied in the 1970s and are consistent with the data obtained on recent beach studies reported in Wymer *et al.* (2004, Table 20, p. 50). However, US EPA (1986) indicates that, whenever possible, the numerical limits adopted by states and tribes should be based on beach-specific log SDs rather than the values 0.40 or 0.70. The value of the log SD used at a beach will depend on the time period over which measurements are obtained. In general, the log SD will be smaller for shorter time periods.

Setting values for the GM and log SD identifies a particular two-parameter lognormal distribution curve. Hence, when US EPA (1986) specifies that the GM of the lognormal distribution equals 33 and the log SD equals 0.4, a specific lognormal distribution curve is being identified as the standard. The beach is not in compliance if either the estimated GM based on *n* measurements exceeds the 50th percentile (median) of that standard distribution or if any of the *n* measurements exceeds the 75th percentile of that standard distribution (or the 82nd, 90th or 95th percentiles as selected).

US EPA (1986) cites day-to-day variability in individual density count measurements as the rationale for setting limits for individual measurements at higher levels than for the GM. However, US EPA (1986), p. 9 considered the GM limits to be most important: 'In deciding whether a beach should be left open, it is the long term geometric mean bacterial density that is of interest.' It appears that the EPA believed at that time that the long-term (30-day) GM limit was protective and the higher limits for individual measurements would protect against unnecessarily closing the beach due to one or a few high measurements.

4.3   Factors in choosing between PC and AC

This section discusses several factors that EPA may choose to consider in deciding whether a PC or AC philosophy, to be described in detail in Sections 4.4 and 4.5, should be adopted by the EPA for the new ambient water quality criteria for bacteria.

4.3.1   Factor 1: Goals and objectives of beach sampling

US EPA (2002, p. 1–1) asks:

Is it safe to swim at local beaches today?
What are the best ways to communicate current water quality conditions to the public?

These questions suggest that accessing the quality of current (today's) conditions of beach water is required. The exposure of swimmers to high levels of fecal bacteria in beach waters must be limited to short time periods. This implies that rapid methods of measuring indicator densities are needed so that samples can be collected, measured and statistically analyzed for decision making as soon as possible after the samples have been collected, preferable on the same day; this would eliminate basing decisions on 30-day means.

4.3.2   Factor 2: Purposes of control charts and acceptance sampling

Natrella (1966, p. 18–1) states that control charts can be used for laboratory applications as a form of statistical test in which the primary objective is to test whether or not a process is in statistical control, i.e. that repeated samples from the process behave as random samples from a stable probability distribution.

Control charts have also been discussed for environmental applications. Gibbons (1994, pp. 160–174) describes how Shewart control charts and cumulative summation (CUSUM) control charts can be used to detect abrupt or gradual changes, respectively, in groundwater contamination from historical levels for individual groundwater wells. The latter application suggests that control chart methods might be used to detect changes from *historical* levels in *E. coli* or enterococci measurements in beach water. However, it should be noted that the current EPA limits shown in Table 4.1 are specific upper-limit values that are not necessarily equal to historical levels at any particular beach. Hence, it appears that US EPA (1986) intended to compare current data with fixed upper limits, not with background, reference, or historical data obtained at the beach. Of course, the fixed upper limits could be based on historical levels if that made sense from a risk perspective.

Consideration should be given to whether the concept of 'statistical control' can actually be applied to bacteria indicator measurements at a beach. These measurements can vary greatly from day to day due to both random and assignable causes. Wymer *et al.* (2004) reported that the GM indicator density among all beach water samples that were collected changed by a factor of 2 (doubling or halving) or more about half the time at each of five studied beaches. Some of the factors that contributed to this variation were:

- spatial factors (distance from shore at which water samples are collected and the location along the shore line);
- temporal factors (the day and time of day water samples were collected);
- the number of swimmers at the beach;
- the amount of rainfall during past 24 hours;
- the amount of sunlight;
- the speed and direction of winds;
- tides;
- water temperature;
- depth in the water column at which water samples are collected.

The use of CCs requires determining 'rational subgroups', i.e., time periods of some specified length (perhaps only a few hours) during which samples are collected to make a decision. For CCs to be useful, the variation of measurements within rational subgroups should include only random (uncontrollable) variation so that a sudden or gradual increase in bacteria levels in the water due to some assignable cause can be distinguished from the random historical variation. If the variation in data within rational subgroups occurs due to both random and assignable causes, then the process is not in statistical control.

Note that AC can be used even when the process is not in statistical control. AC refers to sampling plans and a decision rule that prescribes conditions for acceptance or rejection of a 'lot' (target population). For both CC and AC plans, the beach water target population is the set of all potential water samples of the required size and type that can be collected from the specified beach water area over the time period of interest. Each rational subgroup of data for CCs should be collected from the same beach area. The target population to which the decision applies can be defined to cover any spatial/temporal segment that provides the best data for making correct beach decisions.

The concept of rational subgroups is not needed for the application of AC. However, both CCs and AC require that the target population be defined and that representative samples be collected from the target population. With AC, the variability in measurements within the target population may be caused by both random and assignable causes. But CCs require that the variability within the target population be due only to random causes during some historical time period so that the process is in statistical control, which permits detecting deviations from statistical control at the current time period.

The AC decision rule is a statistical test of a null hypothesis, H_0, about a population parameter of the target population. The null hypothesis is assumed to be true and sampling is conducted to determine if the evidence is sufficiently strong to reject H_0 in favor of a specified alternative hypothesis, H_a. For example, H_0 might be that the true GM (or true arithmetic mean) of the target population is less than or equal to a specified fixed upper limit (threshold value), and H_a might be that the true GM is greater than the fixed upper limit.

Additional discussion of CCs and AC is provided in Sections 4.4 and 4.5, respectively.

4.3.3   Factor 3: Need to control decision error probabilities

Decisions made using CCs or AC will not always be correct. Decision errors can and will occur because the entire target population cannot be measured, and measurements are not perfect. According to the data quality objectives (DQO) process described in US EPA (2006), agreement should be reached – before beach water sampling begins – on the probabilities that can be tolerated of making the two types of decision errors: a Type I decision error (falsely rejecting the null hypothesis) and a Type II decision error (falsely accepting the null hypothesis). These probabilities are selected based on the potential health and monetary costs and consequences of making the two types of decision errors. As the probabilities of making decision errors are set to smaller and smaller levels, more and more samples will need to be collected to achieve those smaller decision error probabilities.

Control charts take into account the probability of making Type I decision errors by specifying the placement of the control lines (see Section 4.4), but little consideration is usually given to controlling the Type II decision error probability.

However, AC plans require specifying both Type I and Type II decision error probabilities when determining the number of samples required for decision making.

4.3.4   Factor 4: Recent statistical publications

Several publications on statistical aspects of assessing water quality standards under Section 303(d) of the Clean Water Act have recently appeared. Smith *et al.* (2001) advocate viewing the water quality assessment process as a statistical decision hypothesis-testing problem wherein both types of decision errors are acknowledged and managed. They studied the performance of the following three statistical approaches that use the number of measurements above the limit to make the decision:

- the EPA 'raw score' approach (US EPA, 1997) in which non-compliance occurs when more than 10 % of the measurements exceed a numeric criteria;

- the binomial test that uses the binomial model to test the null hypothesis that the probability of exceeding the standard is less than or equal to 0.10;

- a Bayesian approach to the binomial test.

The 'raw score' approach is shown to have high Type I decision error probabilities, much higher than the binomial method. The binomial and Bayesian binomial methods allow for controlling both the Type I and Type II error probabilities. They show that low decision error probabilities of the tests cannot be achieved with fewer than 20 samples. Also, they recommend that, at a minimum, the binomial statistical approach to data analysis should be adopted in preference to the 'raw score' approach.

Smith *et al.* (2003) discuss the assessment of water quality using AC methods appropriate for normally distributed data. These methods use the actual values of the measurements rather than the number of measurements than exceed the limit. The performance of the 'raw score', binomial, and AC methods are compared. They conclude that the AC methods discussed are an improvement over the EPA raw score method. They also mention that tolerance intervals could be used to test the null hypothesis (see Section 4.3.5). Equations for computing the number of samples for the AC methods are also provided.

Shabman and Smith (2003) stress that as the amount of monitoring data that can be collected is always limited, it is important to use the most powerful statistical methods available and appropriate to the data. They urge that the process of setting water quality standards proceed with adequate consideration of the (limited) scope of the monitoring program and the statistical procedures that will be used. They state that it makes little sense to define a water quality criterion, as part of a standard, that cannot be measured by the available (or likely to be available) monitoring data. They also note that AC by variables will in most cases require fewer monitoring samples than the EPA 'raw score' or binomial approaches discussed in Smith *et al.* (2001, 2003).

4.3.5　Factor 5: Use of upper tolerance limits rather than CCs or AC

If the revised EPA criteria retain specified percentiles as individual measurement upper limit thresholds, then upper tolerance limits (UTLs) for percentiles rather than CCs or AC may be considered for testing compliance with those individual limits. A UTL on a percentile is equivalent to an upper confidence limit on that percentile. A non-parametric UTL is a particularly easy test to use. One simply determines if the maximum of the n samples collected exceeds the specified limit value. If so, the beach is not in compliance with that limit. The key requirement is that n be large enough. The required n is determined by specifying two inputs: the percentile of interest, e.g., the 75th percentile ($P = 0.75$), and the $100(1 - \alpha)$ percent confidence required in the decision, e.g., 90 % confidence ($\alpha = 0.10$). The required number of samples is then computed as

$$
\begin{aligned}
n &- \frac{\log(\alpha)}{\log(P)} \\
&= \frac{\log(0.10)}{\log(0.75)} = 8.004,
\end{aligned}
\tag{4.1}
$$

which is rounded up to $n = 9$. So, nine representative samples are collected from the target population and if the largest measurement among them exceeds the EPA-specified 75th percentile limit value, then the beach is not in compliance. The largest of the n measurements is, in fact, the UTL. If greater confidence in the decision is required (smaller α) for a constant quantile value P, then n will be larger. For example, if $P = 0.75$ and $\alpha = 0.01$, then $n = 17$. If the data are known to be normally distributed, then the UTL may be computed using the method (equation) prescribed for that distribution, which can be found in, e.g., Hahn and Meeker (1991, p. 60) and Gilbert (1987, equation 11.2, p. 136). If the data are lognormally distributed, the same prescribed method can be used, but applied to the logarithms of the data.

Now, suppose that the true Pth quantile of the target population distribution equals the EPA-specified Pth quantile limit value, which means the beach is just at the point of being not in compliance. Then, assuming independent representative measurements are obtained from the target population, the probability that one or more of n measurements exceeds the limit value, i.e., that the water is declared to not be in compliance, is

$$
1 - P^n.
\tag{4.2}
$$

For example, if $P = 0.75$ and $n = 5$, then $1 - 0.75^5 = 0.763$. That is, the probability that the non-parametric UTL test (comparing the maximum of the five measurements with the 75th percentile limit value) will indicate the beach is not in compliance is 0.76. Hence, equation (4.2) can be used to compute the power (ability) of the non-parametric UTL for individual measurements to detect when the beach is not in compliance. Of course, if the true Pth quantile of the target population distribution is larger than the specified 75th percentile, then the probability that one or more measurements exceeds the specified 75th percentile will be greater than the value computed using (4.2).

Table 4.2 shows values of equation (4.2) for various values of n for the percentiles specified in US EPA (1986), i.e., in Table 4.1. For each value of P in Table 4.2, the true Pth quantile of the target population distribution is assumed to equal the EPA Pth quantile limit value. Table 4.2 makes it clear that using a small n increases the probability that a beach that is not in compliance will fail to be identified. Also, as P increases, the power to detect non-compliance decreases for a given value of n.

4.3.6   Factor 6: Practical constraints

Some practical considerations are:

- As discussed previously, the variation of measurements within 'rational subgroups' should be only random (uncontrollable) variation. Hence, selecting the best 'rational subgroup' will require an understanding of the sources of variation in the measurements at the beach.

- AC methods may be considered more complex to understand and apply than CC methods. The expertise to correctly use AC or even CC methods may not be readily available at all beach locations.

- There are many types of CC and AC plans that could be used. EPA guidance and illustrations may be needed.

Table 4.2   Minimum power (computed using equation (4.2)), of detecting non-compliance of beach water using the US EPA (1986) individual-measurements test of quantiles for values of P and n.

		n		
P	1	5	10	15
0.75	0.25	0.76	0.94	0.99
0.82	0.18	0.63	0.86	0.95
0.90	0.10	0.41	0.65	0.79
0.95	0.05	0.23	0.40	0.54

- Collecting and measuring water samples and making beach closure decisions on the same day are highly desired, regardless of whether CCs or AC is used. This may not be practical at some beaches. New, quick-turnaround measurements of bacterial indicators are needed.

- The need for same-day decisions suggests that field-based laptop computers with user-friendly software could be useful. One possibility is to use the Visual Sample Plan (VSP) software package being developed for the non-statistician. Version 4.0 of VSP and its user's guide (Hassig *et al.*, 2005) can be downloaded free of cost from http://dqo.pnl.gov/. VSP is based on the DQO systematic planning process. Version 4.0 of VSP can be used for the AC designs discussed in this report in Sections 4.5 and 4.6 and for non-parametric upper tolerance limits (equation (4.1)), but would need to be enhanced to include other methods discussed here.

4.4   Control charts

This section provides a brief review and discussion of CCs. The discussion assumes that the count data are lognormally distributed, as does US EPA (1986). In practice, this assumption should be verified for each beach before statistical procedures are used. However, there are reasons why a lognormal distribution might be expected (Ott, 1990).

4.4.1   Shewart control charts

A means Shewart CC provides a running visual graphical record of the mean density counts over time, where each mean is computed using small rational subgroups of data collected periodically from the same beach area. This section provides an example to illustrate the main concepts and computations for a mean CC for beach water sampling when the data are assumed or known to be lognormally distributed. For that case, the logarithms of the density counts are used for all computation. The American Society for Testing and Materials (ASTM, 1976) provides a comprehensive discussion of the construction and use of CCs. Gilbert (1987, pp. 195–202) discusses and illustrates CC for environmental applications. Use of a means CC would imply that the long-term mean is an acceptable basis for ensuring swimmer safety.

Suppose a representative set of n samples (a rational subgroup) is collected from the specified area of beach water during several (k) equal-length time periods. The 'target population' might be the water available for sampling at 9 a.m. each Friday at a depth of 0.3 m from the water surface along a straight line positioned where the water depth (surface to sand) is 0.6 m.

Suppose:

- samples were collected on k consecutive Fridays;

- all water samples are measured for enterococci in units of CFU per 100 ml

- the mean and standard deviation of the logarithms for each of the k sets of n measurements (each rational subgroup) are computed.

The k computed means and standard deviations of the logarithms are used to construct the CC, an example of which is shown in Figure 4.2. The CC in Figure 4.2 consists of a horizontal central line and a horizontal upper control limit line. A lower horizontal limit line is not shown because US EPA (1986) is only concerned with exceeding specified limits.

The central line is computed as the average, $\bar{\bar{x}}$, of the k means of the logarithms. The upper control line is computed as

$$\bar{\bar{x}} + Z \frac{\bar{s}}{\sqrt{n}}, \tag{4.3}$$

where Z is a selected upper percentage point of the normal distribution with mean 0 and standard deviation 1, $\bar{s}$ is the arithmetic mean of the k standard deviations (of the logarithms) for the k rational subgroups (time periods), and n is the number of samples collected and measured for each rational subgroup. In practice, n may differ among rational subgroups.

It is assumed that $\bar{\bar{x}}$ is normally distributed. Note that each SD used in computing $\bar{s}$ is computed using the logarithms of the n enterococci measurements within each rational subgroups. That is, data among the k different rational subgroups are not pooled together to compute a pooled SD based on nk observations. That feature is the key for the CC method to be able to detect when the mean of the enterococci count has increased over historical levels.

Gibbons (1994, p. 161) discusses factors to consider in setting the value of Z. He indicates that the $Z = 3$ might be used if k is large enough that $\bar{\bar{x}}$ and $\bar{s}$ are very precise estimates of the true mean and standard deviation, respectively, of the

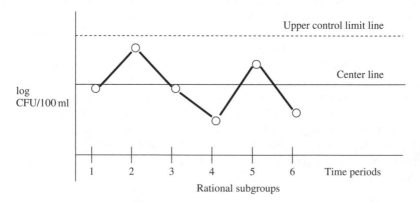

Figure 4.2   Features of a mean Shewart control chart for beach sampling when the density counts are lognormally distributed.

historical process. If $\bar{\bar{x}}$ and $\bar{s}$ are less precisely known, Z might be set to a larger value. However, doing so could lead to less certainty about swimmer safety.

The computed mean of the logarithms of each new set of n enterococci counts obtained each new Friday is plotted on the control chart. ASTM (1976, page 78) states that the process being tracked may not be in statistical control, indicating a change from historical conditions has occurred, if one or more of the means falls above the upper control limit in as few as four or five consecutive rationale subgroups (time periods). In Figure 4.2, none of the means falls outside the control limits, indicating the process is in statistical control thus far. If one or more of the means in future time periods fall outside the limit lines, this is an indication that a new, non-random, source of beach water contamination may be present. In practice, whether one, two or more exceedances of the upper control limit are needed before closing the beach would need to be specified. This decision rule should be selected during the DQO planning process used to determine the sampling plan and the required number of samples while taking into account acceptable Type I and Type II decision error probabilities.

It is important to note that the mean CC compares the current mean enterococci count with historical mean counts, not with a specified EPA upper limit threshold value that may not be related to historical levels.

Also, suppose that $\bar{x}_i$ denotes the mean of the logarithms of the density counts for the ith rational subgroup and we want to make a decision based on only that ith rational subgroup of data. Then using the upper control line of the CC, the decision rule would be to declare that the beach is not in compliance if

$$\bar{x}_i \geq \bar{\bar{x}} + Z\frac{\bar{s}}{\sqrt{n}},$$

i.e., if

$$\bar{x}_i - Z\frac{\bar{s}}{\sqrt{n}} \geq \bar{\bar{x}}. \tag{4.4}$$

Now, the left-hand side of equation (4.4) looks like a lower confidence limit on the mean for the ith time period. However, $\bar{s}$ is computed using historical data, not the n data obtained for the ith rational subgroup for which a decision is needed. If $\bar{s}$ is not representative of the standard deviation for the ith time period, the decision could be in error. AC plans avoid this potential problem by using only the data from the ith rational subgroup to make a decision for that subgroup.

4.4.2   Individual measurement Shewart CCs

One can construct a CC that plots the logarithms of individual measurements as they are obtained. The purpose of such a CC is 'to discover whether the individual observed values differ from the expected value by an amount greater than should be attributed to chance' (ASTM, 1976, p. 97). Again, data are collected in rational subgroups. The center line of the chart is still $\bar{x}$ based on historical data, but the control limit lines are computed differently from those for a mean CC. The method is given in ASTM (1976, pp. 97–98). An individual measurements CC may react more quickly to a sudden large increase in the count of enterococci in beach water compared to the means CC.

4.4.3   Shewart CCs for a standard, not a historical mean

Shewart CCs can also be constructed to discover if observed values of means of logarithms differ from a standard logarithmic value based on either representative prior (historical) data from the process or on a specified desired value. Such a value might be specified by EPA for water quality sampling. However, the CC for a specified desired standard value is set up using a standard value of the standard deviation of logarithms, which would almost invariably be an experience value based on representative historical data (ASTM 1976, p. 76). Note that AC plans do not rely on historical data in making beach closure decisions, only on the data from the latest rational subgroup (time period).

4.5   Acceptance sampling

Rather than using CCs to compare current with historical contamination levels or a specified standard value, one could use the most recent rational subgroup data with an AC plan to test if the threshold limit established for the beach has been exceeded. Wymer *et al.* (2004, p. 66) discuss the AC framework for water quality assessment purposes. A more extensive discussion for industrial applications is in Schilling (1982).

There are two categories of AC plans: *attribute* AC plans, which look at the number of measurements that exceed the limit; and *variables* AC plans, which look at the magnitude of the measurements, usually the mean, in relation to a limit.

4.5.1   Attribute AC plans

Attribute AC plans are non-parametric, i.e., they do not depend on assuming a particular probability distribution for the count data. Also, the target population is assumed to consist of a *finite* number, N (perhaps very large), of water samples that could be obtained during the space-time domain of interest, of which n are actually obtained. There are several attribute AC plans that might be used. Each plan provides a procedure for determining the number, n, of water samples to obtain. A basic attribute AC plan is as follows (Schilling, 1982, Chapter 5):

1. Define p, true proportion of the target population that is 'defective', i.e., the true proportion of the N potential water samples in the target population for

which the density count exceeds the fixed upper bacteria indicator limit for individual measurements.

2. Specify:

   (a) p_2, the unacceptable proportion of the target population of N water samples for which the density count exceeds the upper bacteria limit.

   (b) p_1, the acceptable proportion of the N water samples for which the density count exceeds the upper bacteria limit.

   (c) the null and alternative hypotheses (H_0 and H_a, respectively) to be tested,

$$H_0 : p \leq p_1, \quad H_a : p > p_2.$$

   As H_0 is initially assumed to be true, it is initially assumed that it is safe to swim and the burden of proof is on showing that H_a is true, i.e., that it is not safe to swim.

   (d) α, the producer's risk (probability of making a Type I decision error). This is the tolerable probability of rejecting H_0 when $p \leq p_1$, i.e., the probability we can tolerate that the statistical test used on the count data will incorrectly indicate that it is not safe to swim.

   (e) β, the consumer's risk (probability of making a Type II decision error). This is the tolerable probability that H_0 is accepted when $p > p_2$, i.e., the probability that we can tolerate that the statistical test used on the count data will incorrectly indicate that it is safe to swim.

3. Use the above input parameters in the method by Desu and Raghavarao (1990, pp. 66–67) to compute the number of water samples, n, required as well as the integer c, where $c \geq 0$ is called the 'acceptance number'.

4. Once the n representative density count measurements are obtained, reject H_0 and conclude it is not safe to swim (i.e., accept H_a) if more than c of the n count measurements exceed the fixed upper count density limit.

4.5.2   Attribute compliance sampling

A special case of an attribute AC plan is when $c = 0$, which means that the beach is declared to be unsafe for swimming if one or more individual density count measurements exceed the fixed upper limit count value for individual counts. The philosophy underlying compliance sampling is to determine when an unacceptable proportion of the N target population samples have a density count that exceeds the limit value, rather than distinguishing between an acceptable and an unacceptable proportion of the N samples for which the density count exceeds the limit.

For compliance sampling we specify:

- the number N of potential water samples from the target population;
- p_2, the unacceptable proportion of the N counts in the target population for which the density count exceeds the upper bacteria limit;
- Null and alternative hypotheses,

$$H_0 : p \le p_2, \qquad H_a : p > p_2;$$

- β, the consumer's risk, as defined above;
- $c = 0$.

The above input parameters are then used to compute the number of water samples, n, needed. This can be done using either the method in Schilling (1982, pp. 475–479) or equation (17.8) in Bowen and Bennett (1988, p. 887).

Once the density count measurements are obtained, reject H_0 and conclude it is not safe to swim (i.e., accept H_a) if one or more of the n counts measurements from the lot exceed the fixed upper bacteria limit.

The value of n needed for compliance sampling can be obtained using the compliance sampling module in VSP (Hassig *et al.*, 2005), which uses the method in Bowen and Benett (1988) to compute n. Table 4.3 shows values of n computed using VSP for various values of p_2 and β when the number of potential water samples from the target population is $N = 1000$. For example, from Table 4.3, if $p_2 = 0.10$ and $\beta = 0.25$, then $n = 14$ representative water samples from the target population of $N = 1000$ water samples must be obtained and counted. If all 14 density counts are less than the EPA limit, then we can be $100(1 - \beta)\% = 75\%$ confident that less than $100p_2\% = 10\%$ of the potential $N = 1000$ water density

Table 4.3   Values of n needed for attribute compliance sampling for various combinations of β and p_2 when $N = 1000$.

β	p_2				
	0.40	0.25	0.10	0.05	0.01
0.01	10	16	43	86	368
0.05	6	11	29	57	258
0.10	5	9	22	44	205
0.25	3	5	14	27	129

count measurements exceed the limit. Table 4.3 shows that n can be large if it is important to be highly confident that only a very small proportion of the target population of water samples exceed the EPA limit. For example, if we need to be 99 % confident that less than 5 % of the water sample density counts exceed the EPA limit, then 86 water samples are required and all 86 density counts must be less than the EPA limit.

When the number of units in the target population, N, becomes very large, the number of samples required approaches the number required when non-parametric upper tolerance limits are used to conduct the test (equation (4.1)). VSP can be used to quickly produce tables like Table 4.3 for various values of N.

4.5.3 Variables AC plans

The book by Schilling (1982) describes several variables AC plans. A simple variables AC plan for which the objective is to test whether the mean of the target population of counts (N, assumed to be very large) exceeds a threshold (upper limit value) is as follows:

1. Specify:

   (a) μ_2, the unacceptable mean density count for the target population.

   (b) Null and alternative hypotheses

   $$H_0 : \text{True density count mean} \leq \mu_2,$$
   $$H_a : \text{True density count mean} > \mu_2.$$

   (c) α, the producer's risk (probability of making a Type I decision error). This is the tolerable probability of rejecting H_0 and accepting H_a when H_0 is really true, i.e., the probability we can tolerate that the statistical test (given below) used on the density count data will incorrectly indicate that it is not safe to swim.

   (d) β, the consumer's risk (probability of making a Type II decision error). This is the tolerable probability that H_0 is accepted as being true when H_a is really true, i.e., the probability we can tolerate that the test used on the density count data will incorrectly indicate that it is safe to swim.

   (e) s, the expected standard deviation of representative density count measurements from the target population.

2. Assume that the data, here the logarithms of indicator concentrations, are normally distributed.

3. Compute as follows the number of water samples that must be collected from the target population and counted:

$$n \cong \frac{\left(Z_{1-\alpha} + Z_{1-\beta}\right)^2 s^2}{(\mu_2 - \mu_1)^2} + \frac{Z_{1-\alpha}^2}{2}, \qquad (4.5)$$

where μ_1 is some true mean density count for the target population that, if it should occur, would be acceptable; $Z_{1-\alpha}$ is the $100(1-\alpha)$th percentile of the standard normal distribution (Table A1, p. 254 in Gilbert 1987), with a similar definition for $Z_{1-\beta}$. Equation (4.5) is derived in the Appendix of US EPA (2006),

4. Conduct the following statistical test: Reject H_0 and accept H_a if the lower $100(1-\alpha)\%$ confidence limit on the density count mean exceeds μ_2. Note that H_0 is rejected and H_a is accepted as being true only when the data are very convincing, i.e., only when the lower confidence limit on the mean exceeds μ_2. Hence, the burden of proof is on showing that the true mean density counts of the target population of water samples exceeds the unacceptable mean level μ_2.

Note that one can choose to switch the null and alternative hypotheses to provide more safety for the beach swimmer. That is, one can use

$$H_0 : \text{True density count mean} \geq \mu_2, \qquad H_a : \text{True density count mean} < \mu_2.$$

For this H_0 and H_a, the true mean density count is assumed to equal or exceed μ_2 unless the data are very convincing otherwise, i.e., unless the upper $100(1-\alpha)\%$ confidence limit on the mean is less than μ_2. Hence, in this case the burden of proof is on showing that the true mean is less than the unacceptable mean μ_2. Equation (4.5) is also used to compute the number of required samples, n, for this case of a switched H_0 and H_a.

Table 4.4 provides values of n computed using equation (4.5) for various values of α, β, and μ_1 when $\mu_2 = 1.52$ and $s = 0.40$. (Recall that μ_1, μ_2 and s are in logarithmic units because we are assuming that the log-transformed data have a normal distribution. Under that assumption, the statistical test of H_0 is performed on the logarithms of the density count measurements.) The upper threshold value (unacceptable true mean) is $\mu_2 = 1.52$, which corresponds to the GM threshold of 33 in US EPA (1986). Also, the value $s = 0.40$ is the value of the SD of the logarithms of density counts used in US EPA (1986). The logarithms of μ_1 values 1.40, 1.30, and 1.0 in Table 4.4 are 25, 20, and 10 CFU per 100 ml, respectively. The results in Table 4.4 were obtained using the VSP software (Hassig *et al.*, 2005).

Table 4.4 indicates that fewer than 10 samples have to be collected and measured to make a beach decision if the acceptable true mean density count (on the log scale) is specified to be $\mu_1 = 1.0$ when the unacceptable true mean density count is $\mu_2 = 1.52$ (the EPA limit on the log scale). The required number of samples

Table 4.4 Values of *n* computed using equation (4.5) when $s=0.40$ and $\mu=1.52$.

μ_1	β	$\alpha=0.05$	$\alpha=0.10$	$\alpha=0.20$
1.40	0.05	122	96	70
	0.10	97	74	51
	0.20	71	51	32
1.30	0.05	38	30	21
	0.10	30	23	16
	0.20	22	16	10
1.0	0.05	8	6	5
	0.10	7	5	4
	0.20	6	4	3

decreases as the difference between the unacceptable and acceptable true means increases.

It is important to also consider the sensitivity of the computed value of *n* to the particular value of the standard deviation, *s*, used. The standard deviation value used is assumed to be the true value for the target population, but in reality the true value of *s* is never known. How much does *n* change as the assumed value of *s* changes? This question is addressed in the next section.

4.6 Illustration of a variables AC plan using Visual Sample Plan

Suppose the GM is the statistical parameter of interest and the density counts of enterococci at a freshwater beach of interest are lognormally distributed. The logarithms of the enterococci counts are then normally distributed, so the logarithms of the counts are used in the test. Suppose the EPA has specified that the GM limit value for enterococci is 33 CFU per 100 ml, for which $\log 33=1.52$.

Suppose the user of the VSP software specifies parameter values as follows:

- $\mu_2=1.52$ is the unacceptable true mean of the logarithms of the density counts.

- $\mu_1=1.00$ (determined in consultation with regulators) is the acceptable true mean of the logarithms of the density counts.

- The null and alternative hypotheses are

H_0: True density count mean ≥ 1.52, H_a: True density count mean < 1.52.

This H_0 and H_a put the burden of proof on showing that the true mean of the logarithms of density counts for the target population is less than the unacceptable level 1.52, i.e., that the true GM count is less than the unacceptable level of 33 CFU per 100 ml.

- Consumer's risk: $\alpha = 0.20$ is the tolerable probability that the test will incorrectly indicate that it is safe to swim, i.e., incorrectly indicate that GM < 33 CFU per 100 ml.

- Producer's risk: $\beta = 0.10$ is the tolerable probability that the test will incorrectly indicate that it is not safe to swim, i.e., incorrectly indicate that GM ≥ 33 CFU per 100 ml.

- log $s = 0.40$ (as used in US EPA, 1986).

The VSP software computes that $n = 4$ when these input parameters are used. To use VSP for this illustration, open VSP and click Sampling Goals – Compare Average to a Fixed Threshold – Can assume data will be normally distributed – Ordinary sampling. Enter the parameter values above in the displayed dialog box and click Apply at the bottom of the box to have VSP compute n.

Table 4.5, which was constructed using VSP, shows how the value of n computed using Equation (4.5) changes as the input parameters change. Note that the computed value of n varies from 2 to 12 as the assumed value of the standard deviation of the logarithms of the density counts changes from $s = 0.20$ to $s = 0.80$ when $\alpha = 0.20$, $\beta = 0.10$ and $\mu_1 = 1.00$. This illustrates that it is important to use an accurate value of s when using equation (4.5) to compute n. Table 4.5 also shows how n changes with changes in the other input parameters.

4.7 Discussion

Making good beach closure decisions requires obtaining and appropriately statistically analyzing representative data from the appropriate target population of beach water on a timely basis to minimize the length of time that swimmers are exposed to excessive levels of bacteria. This focus points to the use of AC sampling (test-of-hypothesis) methods using same-day data as opposed to CCs that focus on whether past bacteria levels in the water are continuing unchanged. Beach closure issues require deciding if current conditions, whether or not they are similar to historical levels, are so extreme that beach closure is necessary. Also, using Shewart CCs to make decisions requires that the mean and variance of past data sets are in statistical control, an assumption not needed by AC methods. Although CCs for evaluating whether a specified standard mean value (perhaps based on historical data) is exceeded can also be constructed, the beach closure decision may still depend on the variability observed historically, at least in part.

There is merit in the assertion by Smith *et al.* (2001, 2003) and by Shabman and Smith (2003) that the water quality assessment process should be viewed as a

Table 4.5 Number of samples, n, required for a simple variables AC plan; $\mu_2 = 1.52$.

		Number of samples											
		$\alpha = 0.15$				$\alpha = 0.20$				$\alpha = 0.25$			
μ_1	β	$s=0.2$	$s=0.4$	$s=0.6$	$s=0.8$	$s=0.2$	$s=0.4$	$s=0.6$	$s=0.8$	$s=0.2$	$s=0.4$	$s=0.6$	$s=0.8$
1.368	0.05	13	51	113	200	12	44	97	172	10	38	85	150
	0.10	10	38	85	150	9	32	71	126	7	27	60	107
	0.15	8	31	68	120	7	25	56	99	6	21	46	82
1.246	0.05	5	16	36	63	4	14	31	54	4	12	27	47
	0.10	4	13	27	47	3	10	23	39	3	9	19	33
	0.15	3	10	22	38	3	8	18	31	2	7	15	26
1.125	0.05	3	8	18	30	2	7	15	26	2	6	13	23
	0.10	2	7	13	23	2	5	11	19	2	5	10	16
	0.15	2	5	11	19	2	4	9	15	1	4	7	13
1.003	0.05	2	5	11	18	2	5	9	16	2	4	8	14
	0.10	2	4	8	14	2	4	7	12	1	3	6	10
	0.15	2	4	7	11	1	3	6	9	1	2	5	8
0.882	0.05	2	4	7	12	1	3	6	11	1	3	5	9
	0.10	2	3	6	9	1	3	5	8	1	2	4	7
	0.15	1	3	5	8	1	2	4	6	1	2	3	5
0.760	0.05	2	3	6	9	1	3	5	8	1	2	4	7
	0.10	1	3	4	7	1	2	4	6	1	2	3	5
	0.15	1	2	4	6	1	2	3	5	1	2	3	4

statistical decision hypothesis-testing problem; one in which both types of decision errors (falsely rejecting and falsely accepting H_0) are acknowledged and managed. A concern is the large number of samples that can be required by either CC or AC methods to achieve acceptably small decision error probabilities at beaches when the variability among density counts is high.

One of the challenges faced by EPA in setting new criteria is to consider the limited financial resources, scope of the monitoring program, and the statistical procedures that can or will be used by many communities. The statistical concepts that support hypothesis testing can be difficult to grasp by non-statisticians. The development and use of field-based, user-friendly design, measurement and statistical analysis software for quickly making and documenting cost–benefit trade-offs (cost of samples, benefit of low decision error rates) made during a systematic planning process such as the DQO process are needed. The VSP software, or an enhanced version of it, could be considered for this purpose.

Finally, the use of attribute compliance sampling has an intuitive appeal in that it focuses on determining whether an unacceptable proportion of the target population of water samples exceeds the density count limit value, rather than on distinguishing between an acceptable and an unacceptable proportion of the water volume having concentrations above the limit.

References

American Society for Testing and Materials (1976) *ASTM Manual on Presentation of Data and Control Chart Analysis*. ASTM Special Technical Publication 15D. ASTM, Philadelphia, PA

Bowen, W.M., and Bennett, C.A. (1988) Statistical Methods for Nuclear Material Management, NUREG/CR-4604, PNL-5849. National Technical Information Service, Springfield, VA.

Cabelli, V.J. 1983. Health Effects Criteria for Marine Recreational Waters. US EPA-600/1-80-031. US Environmental Protection Agency, Research Triangle Park, NC.

Desu, M.M., and Raghavarao, D. (1990) *Sample Size Methodology*. Academic Press, New York.

Dufour, A.P. (1984) Health Effects Criteria for Fresh Recreational Waters. EPA 600/1-84-004, U.S. Environmental Protection Agency, Cincinnati, OH.

Gibbons, R.D. (1994) *Statistical Methods for Groundwater Monitoring*. John Wiley & Sons, Inc., New York.

Gilbert. R.O. (1987) *Statistical Methods for Environmental Pollution Monitoring*. John Wiley & Sons, Inc., New York.

Hassig, N.L., Wilson, J.E., Gilbert, R.O., Pulsipher, S.A., and Nuffer, L.L. (2005) *Visual Sample Plan Version 4.0 User's Guide*, PNNL-15247. Pacific Northwest National Laboratory, Richland, WA. Available free at http://dqo.pnl.gov/vsp.

Hahn, G.J., and W.Q. Meeker, (1991) *Statistical Intervals*. John Wiley & Sons, Inc., New York.

Natrella, M.G. 1966. *Experimental Statistics*. National Bureau of Standards Handbook 91. John Wiley & Sons, Inc., New York.

Ott, W.R. (1990) A physical explanation of the lognormality of pollutant concentrations. *Journal of the Air and Waste Management Association*, 40: 1378–1383.

Schilling, E.G. (1982) *Acceptance Sampling in Quality Control*. Marcel Dekker, New York.

Shabman, L., and Smith, E. (2003) Implications of applying statistically based procedures for water quality assessment. *Journal of Water Resources Planning and Management*, July/August: 330–336.

Smith, E.P., Ye, D., Hughes, C., and Shabman, L. (2001) Statistical assessment of violations of water quality standards under Section 303(d) of the Clean Water Act. *Environmental Science and Technology*, 35(3): 606–612.

Smith, E.P., Zahran, A., Mahmoud, M., and Ye, K. (2003) Evaluation of water quality using acceptance sampling by variables. *Enironmetrics*, 14: 373–386.

US Environmental Protection Agency (1986) Ambient water quality criteria for bacteria-1986 (EPA440/5-84-002). US Environmental Protection Agency, Office of Water Regulations and Standards, Criteria and Standards Division, Washington, DC. January 1986.

US Environmental Protection Agency (1997) Guidelines for preparation of the Comprehensive State Water Quality Assessment. Office of Water, US EPA, Washington, DC.

US Environmental Protection Agency (1999) EPA action plan for beaches and recreational waters (EPA/600/R-98/079). US EPA, Washington, DC.

US Environmental Protection Agency (2002) Time-relevant beach and recreational water quality monitoring and reporting. (EPA/625/R-02/017). EMPACT, US EPA, Office of Research and Development, National Risk Management Research Laboratory, Cincinnati, OH, October.

US Environmental Protection Agency (2006) Guidance on systematic planning using the data quality objectives process (EPA QA/G-4, EPA/240/B-06/001). US EPA, Office of Environmental Information, Washington, DC, February.

Wymer, L.J., Dufour, A.P., Brenner, K.P., Martinson, J.W., Stutts, W.R., and Schaub, S.A. (2004) The EMPACT Beaches Project: Results from a study on microbiological monitoring in recreational waters. US EPA, Office of Research and Development, National Exposure Research Laboratory, Microbiological and Chemical Exposure Assessment Research Division, Cincinnati, OH.

5

Sampling recreational waters

A.H. El-Shaarawi and S.R. Esterby

National Water Research Institute, Environment Canada, Burlington, Ontario, Canada, and Mathematics, Statistics and Physics, Irving K. Barber School of Arts and Sciences, University of British Columbia Okanagan, Canada

5.1 Introduction

This chapter presents a number of statistical techniques suitable for use in the assessment of recreational water quality. The primary focus is on sampling designs for measuring indicator bacteria concentrations in recreational water and inferential methods for the estimation of parameters, including design-based, model-based, and a combination of design-based and model-based methods. Typically, the concentration of an indicator microbiological organism is measured at several locations within the recreational area and summary statistics are computed from the data. Based on the values of these statistics, a decision is made about the safety of the water for recreational activities. It is critical to collect data in such a way that numbers used in the decision-making process are as precise as constraints, such as cost, permit and represent accurately the state of the water body. Thus, sampling designs that include an adequate number of sampling points and account for known sources of variability, and protocols which ensure high-quality data are essential to an effective decision-making process. This chapter summarizes some important statistical techniques that are helpful in selecting a representative and efficient

Statistical Framework for Recreational Water Quality Criteria and Monitoring   Edited by Larry J. Wymer
© 2007 John Wiley & Sons, Ltd

sampling plan for recreational waters. The purpose of the data collection could be for research to determine sources and magnitude of variability in indicator bacteria concentrations, regular monitoring or event-driven sampling.

The chapter is organized as follows. Section 5.2 begins with a brief description of the concepts of population and sample and gives some useful definitions that will be needed throughout. Important concepts related to data quality and to the effective number of observations in the presence of autocorrelation are also considered. Section 5.3 is concerned with components of variability in microbiological data; the nature of the components, modeling variability in terms of probability distributions appropriate for this type of data, and a brief example of inference for population parameters. In Section 5.4, several design-based sampling techniques are presented: simple random sampling, stratified random sampling, systematic sampling, ranked set sampling, and composite sampling. A very brief comment on model-based sampling is presented in Section 5.5, and some concluding remarks are made in Section 5.6.

5.2   Data quality

5.2.1   Populations and samples: some definitions

Consider the general setting where statistics calculated from measurements on water samples are compared with specific criteria as the method of determining whether a recreational area is safe or of attaching some level of risk for illness. Whether it is explicitly stated or implicit, the criteria are linked to the presence or level of pathogenic bacteria in the water contacted by the individual. The link between presence or level of bacteria and the risk of illness, and the choice of the particular criteria, are not topics of this chapter. However, they must be considered in practice when deciding how to sample water for the assessment of safety. Here attention is restricted to the following: the problem of defining the particular volume of water in terms of spatial boundaries and temporal intervals, considerations of the distribution of bacteria within this volume and factors which affect how the bacteria are distributed, and sampling methods which allow us to take water samples in a well-defined way. The latter includes statistical sampling designs and constructs, such as grids and transects, which permit us to go from populations to samples.

Several definitions are required prior to discussing the selection of representative samples that will produce precise information about the quality of recreational water. First, the *target population* is the area for which safety for recreational activities is to be assessed. Some method for defining all possible sampling points within the target population must be determined. A map of the area is frequently found appropriate to represent this population. Clearly the population is continuous and thus consists of a very large number of sampling points and can be viewed as an infinite population.

Second, the *sampled population* is the area for which the assessment can be legitimately made from the sample. The main source of discrepancy between the target and sampled populations is the inadequacy of the map used to draw the sampling points. This discrepancy is a serious source of bias.

Third, *subpopulations* or *strata* are obtained when the target population is heterogeneous with respect to bacterial concentrations and knowledge exists that allows the total area to be divided into a number of subpopulations or homogeneous areas. The pollution levels need to be estimated for each stratum and compared. An overall estimate for level in the population is obtained by combining the estimates of the strata levels.

Finally, the *sample* consists of the points at which microbiological measurements are to be made, and the sample design refers to the technique used to select the sample.

5.2.2   Characteristics of data quality

Data are collected for specific goals and thus the achievement of the goals within the resources available is a measure of data quality. For example, the 1986 US Environmental Protection Agency criterion for recreational waters, originally formulated in 1968, recommends the use of fecal coliform as the water quality indicator organism, with concentrations to be determined by multiple-tube fermentation or membrane filter procedures (US EPA 1986). The EPA decision rule states that the coliform content of primary contact recreation waters shall not exceed a log mean of 200 per 100 ml nor shall more than 10 % of total samples collected during any 30-day period exceed 400 per 100 ml. This is a specific example of the general setting referred to above. The collection of data from a recreational area will be used to make a decision about water safety based on computing those two statistics and comparing them to EPA limits. However, there are uncertainties in observed values of the statistics due to known and unknown sources of variation. A good data set is the one that has been collected under a sampling plan that efficiently uses known information and guards against unknown sources of variability in setting the sampling design. The three important characteristics of data of good quality are:

1. *The absence of systematic bias.* Bias is a persisting, one-directional error, which could be caused by an inadequate sampling design, analytical methods, field protocols, or errors in data processing and reporting. To illustrate, suppose that one decides to estimate the average density of an indicator organism in a recreational area by examining water samples taken only at the chest depth. Thus the sampling area does not include part of the target area, and this will result in bias. If the bacterial density declines with depth, the computed average will underestimate the target area average resulting in a negative bias. Here the direction of the bias is known. On the other hand, if a faulty microbiological analytical method is used in the analysis of the water samples, the direction of the bias is likely to be unknown. It is clear in both cases that making more measurements under the same conditions will not alleviate the bias problem.

2. *High precision.* This refers to the desire to estimate the quantities of interest with the highest possible precision within the resources available. Usually the variance and sometimes the mean square error are used as criteria to measure the precision of the estimator. A best sampling technique is one that minimizes the chosen criterion. Depending on the population sampled, the more observations made, the more precise is the estimator. The use of known population features at the data collection and analysis stages increases the precision of the estimators, and how this is achieved will be discussed in Section 5.3. It is important that design allows for the estimation of the precision of the estimator from the collected data.

3. *Generality of the results.* Studies are frequently conducted where the scope is limited but the objective requires the findings to be applicable on a broader scale. For example, a design for monitoring recreational beaches in a country is usually established based on the results gained from the study of a limited number of carefully selected beaches. A good design will provide data that permit extraction of information that is applicable to a broader population beyond the one inherent to the data generated.

5.2.3   Effective number of observations: design implications

Neighboring locations within a recreational area are expected to be similar, which means that their observations are positively correlated. This correlation is expected to decline as the distance between the locations increases. Analogously, observations at the same location are usually positively correlated when they are close together in time. The effect of serial correlation on the precision of the estimator of the population mean was considered by Bayley and Hammersley (1946). Here we consider the simple case of estimating the population mean. Let $x_1, x_2, \ldots, x_n$ be a realization from an equally spaced stationary time series process with mean μ and variance σ^2, and where the correlation between x_i and x_{i+d} is ρ^d, with d being the time between the two observations and $0 \leq \rho \leq 1$. This also applies to spatial sampling in one dimension. The average $\bar{x}$ is a natural estimator of μ, and its variance is given by

$$\text{Var}(\bar{x}) = \frac{\sigma^2}{n}\left(1 + 2\sum_{i=1}^{n-1}\left(1 - \frac{i}{n}\right)\rho^{di}\right). \tag{5.1}$$

This expression shows that the variance increases as ρ increases. This means that the highest precision occurs when the observations are uncorrelated and the precision may be improved by increasing the distance between the sampling points.

Equation (5.1) may be used to determine how many uncorrelated observations, denoted by n_E, are needed to yield the same variance for the estimated mean as

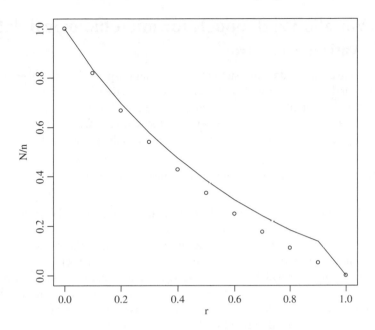

Figure 5.1   The ratio n_E/n versus the correlation coefficient ρ for n, the number of correlated observations, equal to 10 and equal variances for both sample means.

obtained from n correlated observations. By equating the variance of $\bar{x}$ for n_E uncorrelated observations with the expression in equation (5.1), we obtain

$$n_E = \frac{n}{1 + 2\sum_{i=1}^{n-1}(1 - i/n)\rho^{di}}. \tag{5.2}$$

A simpler expression can be obtained by noting that for large n, (5.2) is approximately (El-Shaarawi and Damsleth, 1988)

$$n_E = \frac{(1 - \rho^d)n}{1 + \rho^d}. \tag{5.3}$$

In Figure 5.1, the ratio n_E/n is plotted against ρ using equation (5.3) (circles) and equation (5.2) (continuous line) for $d = 1$, the case where observations are one time unit apart. It shows that, even for n as small as 10, the simple expression provides a good approximation. The value of n_E can be computed for various values of n and ρ. Under positive correlation, more observations are required to achieve the same precision as obtained when correlation is absent. For example, for serial correlation as small as $\rho = 0.10$, 100 observations yield the same precision in the estimation of the population mean as that obtained from only 82 observations when $\rho = 0$. The effective number of observations would be 82 in this case.

5.3 Variation and models for microbiological data

5.3.1 Variation and error

Microbiological count data are subject to two main types of variability: sampling variability and natural variability. Suppose the objective is to estimate the mean μ of the target population using the sample mean $\bar{x}$ obtained from measurements made at n locations. The difference between the sample mean and the population mean μ is just $\bar{x} - \mu$, the error of the estimation, which can be written as

$$\bar{x} - \mu = \text{sampling error} + \text{measurement error}.$$

Each of the components could be subject to bias, as discussed above, but here we concentrate on the random causes of error.

Sampling error or *variability* is the amount by which the estimate from the sample would deviate from the true value μ when the measurements are exact. It is that part of the error which is due to the fact we are able to make measurements on part of the target population. The uncertainty due to this variability depends on the sampling design and the number of locations used. It can usually be estimated from the data, as will be described in the next section.

Measurement error is caused by the error and variability associated with an individual measurement. For example, repeated measurements at the same location or repeated microbiological analyses of the same water sample will yield variable results.

The classification of inferential methods into design-based and model-based techniques is associated with these two error components. However, it is important to realize that in real situations the combination of both techniques is usually necessary for the construction of good designs and efficient estimation. To achieve this, the target recreational water population is viewed as a superpopulation. Within this superpopulation or infinite population is the finite population of N locations. The finite population is then the frame from which a sample of n locations is selected for making the measurements. A reasonable approach is to divide the superpopulation into a large number of N cells and to assume that water quality within each cell is homogeneous. The central points of these cells are assumed to represent the elements of the finite population. A random selection of n points out of N possible locations will represent the elements of the sample.

5.3.2 Accounting for known patterns of variation

Let $X_1, X_2, \ldots, X_N$ be the indicator organism concentrations at the N cells that represent the finite population and $x_1, x_2, \ldots, x_n$ be the outcome obtained from the sample of size n. In design-based sampling, the $X_1, X_2, \ldots, X_N$ are assumed to be fixed but unknown constants and the error incurred is simply due to the fact that only some of the sampling locations are selected. The finite-population characteristics are unknown constants that are to be estimated from the sample. For example, the population mean and variance are given by

$$\bar{X} = \frac{1}{N}\sum_{i=1}^{N} X_i \quad \text{and} \quad S^2 = \frac{1}{N}\sum_{i=1}^{N}(X_i - \bar{X})^2.$$

The corresponding sample estimates of these quantities are

$$\bar{x} = \frac{1}{n}\sum_{i=1}^{n} x_i \quad \text{and} \quad s^2 = \frac{1}{n}\sum_{i=1}^{n}(x_i - \bar{x})^2.$$

Under the finite-population assumptions, the precision of these estimates is determined by the sampling technique used to draw the sample, as will be shown in the next section.

In a model-based approach, to allow for variability due to measurement error, distributional assumptions about the variables being measured are introduced. The elements $X_1, X_2, \ldots, X_N$ of the finite population are assumed to be drawn from a superpopulation with mean, variance, and covariance given by

$$\mu = E(X_i), \qquad \sigma^2 = E(X_i - \mu)^2, \qquad \rho_d \sigma^2 = E(X_i - \mu)(X_j - \mu), \qquad (5.4)$$

where d is the distance between locations i and j. It is apparent that the model-based design does not require that the elements of the sample $x_1, x_2, \ldots, x_n$ be selected at random, but in this case it will not be possible to evaluate the estimator's total uncertainty. The efficiency can be determined under assumptions (5.4) without referring to the sample design.

In the hybrid design/model approach, both the particular sampling technique to be used in selecting the sample and distributional assumptions about the elements of the finite population, such as those given in (5.4), need to be used to determine the sample design. In addition to accounting for measurement error, there may be explanatory variables which could be included through the model-based approach to improve the information content of the sample.

5.3.3 Models for bacterial counts

The lognormal $LN(\mu, \sigma^2)$, Poisson $P(\lambda)$, and negative binomial $NB(\nu, \mu)$ probability distributions are widely used as models to describe the variability in microbiological data. The parameters of these distributions, μ, σ^2, λ, and ν, will generally be estimated from the sample data. For applications of the Poisson and negative binomial distributions, see Fisher *et al.* (1922), Fisher (1941), El-Shaarawi *et al.* (1981) and El-Shaarawi (2003a); for the lognormal distribution, see Aitchison and Brown (1981) and El-Shaarawi (2003b). The lognormal distribution is particularly popular among environmental agencies in developed countries as a model in the setting of microbiological standards and guidelines. For example, the 1992 Health

Table 5.1   Characteristics of the normal distribution and of distributions frequently used in the analysis of bacteriological data.

	Normal	Lognormal	Poisson	Negative binomial
Density $f(x)$	$\dfrac{e^{-\frac{(x-\mu)^2}{2\sigma^2}}}{\sqrt{2\pi}\sigma}$	$\dfrac{e^{-\frac{(\ln(x)-\eta)^2}{2\sigma^2}}}{\sqrt{2\pi\sigma^2 x^2}}$	$\dfrac{e^{-\mu}\mu^x}{x!}$	$\dfrac{\Gamma(\nu+x)}{\Gamma(\nu)\Gamma(x)}\dfrac{\nu^\nu\mu^x}{(\nu+\mu)^{\nu+x}}$
Mean	μ	$e^{\eta+\sigma^2/2}$	μ	μ
Variance	σ^2	$\mu^2(e^{\sigma^2}-1)$	μ	$\mu+\dfrac{\mu^2}{\nu}$
Skewness	0	$(e^{\sigma^2}+2)\sqrt{e^{\sigma^2}-1}$	$\mu^{-1/2}$	$\dfrac{\nu+2\mu}{\sqrt{\nu\mu(\nu+\mu)}}$
Kurtosis	3	$e^{4\sigma^2}+2e^{3\sigma^2}+3e^{2\sigma^2}-3$	$3+\mu^{-1}$	$3+6\nu^{-1}+\left(\mu+\dfrac{\mu^2}{\nu}\right)^{-2}$

Canada guideline for recreational water sets the limit for *E. coli* in terms of the geometric mean, which is justified by the assumption of lognormality (El-Shaarawi and Marsalek, 1999). Table 5.1 gives the functional forms of these distributions and a summary of their characteristics. The normal distribution is included because of its common use in statistical inference. It should be noted that the other three distributions in the table are skewed and have higher kurtosis than the normal distribution.

Figure 5.2 shows the density of the lognormal, negative binomial, and Poisson plus added zeros when the parameters of the three distributions are specified so that they have common means and variances. This shows that the lognormal distribution is the most skewed, followed by the negative binomial, and then by the Poisson plus zeros. Since these distributions have only two parameters, the maximized likelihood is the most appropriate method for discriminating among them.

5.3.4   Estimation and inference based on the models

Given a sample $(x_1,\ldots,x_n)$ of n observations from $LN(\mu,\sigma^2)$ with $x_1 = \log(y_1),\ldots,x_n=\log(y_n)$, then the sample mean $\bar{x}=\sum_{i=1}^n x_i/n$ and variance $s^2 = \sum_{i=1}^n (x_i-\bar{x})^2/(n-1)$ are the unbiased estimators of μ and σ^2, respectively (Lehmann, 1983). Exact inferences about μ and σ^2 are available using the Student $t_{(n-1)}$ and $\chi^2_{(n-1)}$ distributions. For example, the $100(1-2p)\%$ confidence intervals for μ and σ^2 are

$$\left(\bar{x}-t_{1-p,(n-1)}\frac{s}{\sqrt{n}},\bar{x}+t_{1-p,(n-1)}\frac{s}{\sqrt{n}}\right) \quad \text{and} \quad \left(\frac{(n-1)s^2}{\chi^2_{1-p,(n-1)}},\frac{(n-1)s^2}{\chi^2_{p,(n-1)}}\right) \qquad (5.5)$$

where $t_{p,(n-1)}$ and $\chi^2_{p,(n-1)}$ denote the $100p$th percentiles of the $t_{(n-1)}$ and $\chi^2_{(n-1)}$ distributions, respectively.

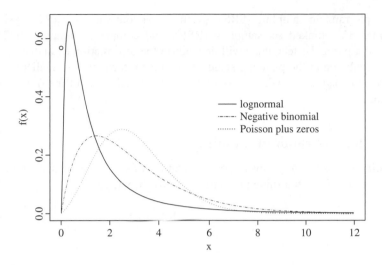

Figure 5.2 The shape of the lognormal, Poisson plus added zeros, and negative binomial distributions when their means and variances are equal.

Inferences are frequently required for a function of the form:

$$\eta(b) = e^{\mu + b\sigma^2}, \tag{5.6}$$

where b is a known constant. This function gives the mode, median, and mean of the lognormal distribution for $b = -1, 0, 0.5$ respectively. The maximum likelihood estimate of $\eta(b)$ is $\hat{\eta}(b) = e^{\bar{x} + bs^2}$, which is biased upwards. For large n the bias can be ignored, especially when s is small. El-Shaarawi (1989) gave the $100(1 - 2p)\,\%$ asymptotic confidence interval for the mean as

$$\hat{\eta}(b) e^{\pm z_{1-p}\sqrt{(s^2 + 2b^2 s^4)/n}}, \tag{5.7}$$

where z_p is the $100p$th percentage point of the standard normal distribution. El-Shaarawi and Lin (2006) showed that replacing z_p by the $100p$th percentage point of Student t distribution with $(n-1)$ degrees of freedom yields a very precise confidence interval for the lognormal distribution for all small samples. It should be mentioned also that the jackknife and bootstrap methods are also available for bias-corrected estimator.

5.4 Design-based sampling techniques

Five basic probability sampling selection techniques are briefly discussed in this section. Each of these techniques will select a sample of n points out of the N points that constitute the finite population with a well defined probability. These are: simple random sampling (SRS), stratified random sampling (ST), systematic sampling (SY), ranked set sampling (RSS), and composite sampling (CS). The choice of a particular technique will depend on the information available about the target population at the planning stage. If no prior information is available, it may be useful to conduct a small pilot study to guide the selection of the final sampling selection.

5.4.1   Simple random sampling

According to this technique every possible sample of size n has the same probability π_s of selection from the finite population of size N, where

$$\pi_S = \frac{1}{\binom{N}{n}} = \frac{n!(N-n)!}{N!},$$

and $b! = 1 \times 2 \times \ldots \times b$. Figure 5.3 provides an example of a rectangular recreational area with $N = 80$ sampling points and shows a simple random sample of size $n = 10$. The chosen sample is just one of more than 1.5 billion possible samples under SRS. The characteristics of any estimator computed from the sample are assessed from its distribution over all the possible samples. Based on this, Cochran

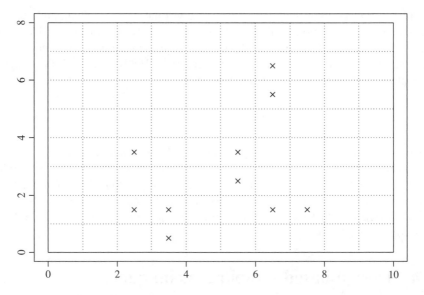

Figure 5.3   A realization of a simple random sample of size $n = 10$ from a population of size $N = 80$ cells obtained by a grid placed on the area of interest.

(1977) showed that the sample mean $\bar{x}$ is an unbiased estimator of the population average $\bar{X}$ with estimated variance

$$\text{Var}(\bar{x}) = s_{\bar{x}}^2 = \frac{s^2}{n}(1 - f),$$

where

$$s^2 = \frac{\sum_{i=1}^{n}(x_i - \bar{x})^2}{n - 1}, \qquad f = \frac{n}{N}.$$

Invoking the central limit theorem, the approximate confidence limits for $\bar{X}$ are $\bar{x} \pm z_\alpha s_{\bar{x}}$, where z_α is $1 - 2\alpha$ quantile of the standard normal distribution. It is usually the case that real sampling populations are large, hence f is small and can be regarded as 0. Therefore the width of the confidence interval of the mean depends mainly on n for a given population. Given the availability of a rough estimate for the variance, an approximate estimate of n can be determined that will lead to an interval of specified width.

To illustrate the details, let us consider a manageable example of a population with $N = 8$ sampling points and suppose that the concentrations of the indicator organisms at these locations are 13, 19, 17, 18, 16, 11, 10, 20. Then the population mean is $\bar{X} = 15.5$ and the population variance is $S^2 = 12.25$. Using the software package R, the random sample 10, 17, 19 of size $n = 3$ is selected. This sample gave $\bar{x} = 15.33$ and $s^2 = 22.33$, so that $\text{Var}(\bar{x}) = 4.65$ and the approximate confidence interval for $\bar{X}$ is $(11.1, 19.6)$.

When additional information is available about the population it should be used to obtain a more efficient sampling design. To compare sampling designs it is necessary to introduce a definition of efficiency. The most efficient design is the one that estimates the quantity of interest with maximum precision at a fixed cost and without bias. Suppose that the population mean μ is the quantity of interest and let $\hat{\mu}$ be the estimator based on the most efficient sampling design. Then the estimator is (1) unbiased when its expected value is μ, that is $E(\hat{\mu}) = \mu$, and (2) most efficient when it has the smallest variance among all possible estimators.

5.4.2 Stratified random sampling

Stratification is a device that may result in a gain in precision over simple random sampling when supplementary information is used to group the target population into a number of subpopulations (or strata). The method involves taking a simple random sample from each of these strata. Stratification is effective when variation within the strata is much smaller than variation between them. Thus maximum precision is achieved when the strata averages are as different as possible, and their variances are as small as possible. For example, depth is an important variable to take into account when sampling beaches (Wymer *et al.*, 2005). One strategy would be to divide a beach swimming area into three zones (ankle-depth, knee-depth, and chest-depth zones) and take a random sample from each zone since the bacterial densities would be expected to be more similar within zones than between zones.

Let the N locations be divided into k strata such that N_i locations are in the ith stratum and $N = \sum_{i=1}^{k} N_i$. The objectives are to estimate the average of the whole population and the average of each stratum by taking a sample of size n, where $n = \sum_{i=1}^{k} n_i$ and n_i is the number of locations allotted to the ith stratum. There are questions of how to allocate n among the different stratum and how to combine the different stratum estimates in the estimation of the population average. Optimal allocation depends on the variation in the N_i and the differences between the within-stratum variances. Estimation of the mean of ith population stratum is simply the mean of the n_i sampling points allocated to the ith stratum, while the population mean is the weighted mean of the strata means with weight proportional to the N_i.

5.4.3   Systematic sampling

This method is very popular because it is simple to execute in the field and because of the ability to spread the sample more evenly over the target population. The starting point is to divide the target population into k groups each of size n and randomly select one of these groups as the required sample. Thus a systematic sample is equivalent to a cluster sample, where knowledge of the order of sampling points in the population is used in forming the groups and one group is selected as the sample. Equivalently, the sampling points of the population are grouped in n classes each of size k and the sampling points within classes are sequentially ordered with the k sampling points in the ith $(i = 1, \ldots, n)$ class numbered as $(i-1)k+1, (i-1)k+2, \ldots, ik$. The systematic sample will then consist of the points $j, j+k, j+2k, \ldots, j+(n-1)k$, where j is a number selected at random from the numbers 1 to k.

The following example illustrates the process of selecting a systematic sample of three locations from a recreational area that is assumed to be represented by a finite population of nine locations. Suppose the indicator counts at these locations are

$$28, 11, 19, 25, 27, 30, 17, 20, 14.$$

Thus the true population mean $\bar{X} = 21.22$ is the quantity which is to be estimated from a systematic sample of size $n = 3$. The columns of Table 5.2 show the three systematic samples, if the nine sampling points are divided into three classes as shown in the rows of the table.

One obvious difference between SRS, ST, and SY is the difference in the number of possible samples under each of these techniques. In the example above, we have only three possible SY samples, thus the probability of selecting any one of the samples is 1/3. This can be contrasted with 84 possible sample selections under SRS with a selection probability 1/84, and 27 possible samples under ST with a selection probability of 1/27. The restriction to the sample selections as a result of knowledge about the population structure in the cases of SY and ST reduces the number of possible samples that can be selected.

Table 5.2 All possible systematic samples of size 3.

	Sample I	Sample II	Sample III
Class I	28	11	19
Class II	25	27	30
Class III	17	20	14

Cochran (1977) has shown that the sample mean is an unbiased estimator for the population mean. The efficiency of this estimator is dependent in the structure of the population. For some populations SYS is less efficient than SRS and for other populations SYS is very efficient. The reader should consult Cochran (1977) for more details. Here we just state that SYS will be very precise when the variances within the samples are maximized, which is equivalent to minimizing the between-sample variances. One such situation is when the elements of the population are actually ordered.

Another serious problem is the impossibility of estimating the variance of the sampling mean because SYS is a single sample out of the k possible samples. Further assumptions about the finite population are required in order to be able to estimate the population variance.

5.4.4 Composite sampling

The use of composite sampling to collect environmental data is increasing. It has the advantage of increasing the spatial and temporal coverage of the target population without increasing the laboratory work. It is sometimes the only method available to collect environmental data, as in the collection of precipitation samples (rain and snow). It is a technique where multiple temporally or spatially discrete samples are combined, thoroughly homogenized, and treated as a single sample. This method is not new; early examples include testing US servicemen for syphilis in the Second World War (Dorfman, 1943) and group testing of the prevalence of insects carrying a plant virus (Watson, 1936). Barnett and Bown (2002) described its use in setting standards for contaminated sites.

CS is expected to reduce variability due to the combination of samples with extreme values. When it is desirable to identify critical points within a sampling region, as in the case of extreme water quality, it may be necessary to allow the examination of the discrete samples that constitute extreme composite samples. This would require the preservation of the discrete samples for future analysis. This may, however, cause difficulties in the interpretation because the results obtained can be confounded with changes induced by the preservation process and its duration. It is clear that the suitability of CS is dependent on the objectives of data collection and on the ratio of the laboratory cost to the cost of field sampling. If the ratio is substantially larger than unity, CS will be appropriate. The objectives here are to describe the use and efficiency of CS in testing for the presence of microbiological contamination and in estimating the average concentration of the indicator organism in recreational waters.

Composite sampling will be effective when testing for a rare organism in water, particularly when there are serious concerns that the organism be detected. Consider a recreational area with N sampling locations but where it is possible only to perform the microbiological analyses on n samples ($n < N$). This can be done in two ways:

1. Take a water sample from each of n randomly selected locations and conduct microbiological analysis on each of the n water samples separately, thus n microbiological tests are required.

2. Pool water samples from c locations and test them together. If the test is negative, a single test is sufficient for the c locations. If the test is positive, each of the c samples must be tested separately. Do this for each of the n/c composite samples.

To study the relative efficiency of pooling samples, let p be the probability that the test yields a positive result for any one of the locations and assume that locations are stochastically independent. Then the probability that the test on a composite sample, obtained from pooling sample from c locations, will be positive is

$$p_c = 1 - (1 - p)^c,$$

and thus the expected value of the number of tests under plan 2 is

$$E(\text{tests}) = \left(p_c + \frac{1}{c}\right) n.$$

The number c is a design parameter which needs to be selected by the microbiologist using prior knowledge about the value of p. Indeed, it can be shown (Feller, 1968) that when $c \approx 1/\sqrt{p}$ the minimum expected number of tests under CS is approximately $2n\sqrt{p}$. It is then clear that pooling is more efficient than single testing when $0 \leq p < 1/4$.

The discussion here is restricted to estimation by the method of moments (MM). This is a non-parametric method of estimation since it does not require the specification of the form of the probability distribution for the concentration of the indicator organism. Let μ, σ^2, μ_3 and $\mu_4 = \beta\sigma^4$ be the mean, variance, and the third and fourth central moments of the distribution, respectively. Let $x_{i1}, \ldots, x_{ic}$ be the concentrations of the c discrete samples that forms the ith composite sample ($i = 1, \ldots, n$). Then the concentration in the sample is $y_i = \sum_{j=1}^{c} x_{ic}/c$. Thus the MM estimators of μ and σ^2 are given respectively by

$$\bar{y} = \sum_{i=1}^{n} \frac{y_i}{n}, \qquad s^2 = \frac{c}{n-1} \sum (y_i - \bar{y})^2.$$

To determine the distributional properties of these estimators, it is necessary to evaluate their moments, particularly their variances which measure the precision.

Consider first the properties of the MM estimator of μ. It can be shown using the moment generating function that the distribution of $\bar{y}$ based on n composite samples is exactly the same as that obtained when the concentrations of nc discrete samples are determined. In particular, the expressions (Cramér, 1946) for the first four central moments of the distribution of $\bar{y}$ are:

$$E(\bar{y}) = \mu, \qquad \text{Var}(\bar{y}) = \frac{\sigma^2}{nc},$$

$$\mu_3(\bar{y}) = \frac{\mu_3}{(nc)^2}, \qquad \mu_4(\bar{y}) = \frac{\sigma^4(\beta + 3nc - 3)}{(nc)^3}.$$

These show that $\bar{y}$ is an unbiased estimator of μ and the higher moments decline with either increasing n or increasing c. An immediate consequence of this is that, for fixed n, the distribution of $\bar{y}$ converges faster to the normal distribution as c increases. Although we might ask if this means that we can take a single composite sample with $c = n$, clearly this does not sense since then it would not be possible to estimate σ^2, which is needed for the precision of the estimator of μ.

Clearly, the properties of the MM estimator of σ^2 are important. It can be shown that s^2 is an unbiased estimator of σ^2 regardless of the values taken by c and n for $n > 1$. The variance of this estimator is given by

$$\text{Var}(s^2) = \frac{\sigma^4 \{(n-1)\beta + (2cn - 3n + 3)\}}{n(n-1)c} = \frac{2\sigma^4}{n-1} \left\{ \frac{(\beta - 3)(n-1)}{2nc} + 1 \right\}.$$

This expression depends on Fisher's measure of kurtosis $\beta - 3$, which is zero for a normal distribution and may take negative and positive values for other distributions. For a normal population the variance of s^2 is independent of c, the number of individual samples that constitutes a composite sample. The expression also shows that the factor by which this variance is influenced by non-normality is almost independent of n and it depends mainly on c. The variance decreases as c increases, and for distributions with negative kurtosis CS is more efficient than CS under normality. The opposite is true for distributions with positive kurtosis. It should be noted that the frequently encountered distributions (Table 5.1) in the analysis of microbiological data sets have positive kurtosis and thus the MM estimator of σ^2 under CS is less efficient than the one based on a simple random samples of size n.

5.4.5 Ranked set sampling

This method, introduced by McIntyre (1952), combines random sampling and the ability to rank the sampling units with respect to the characteristic of interest by any means other than actual measurements of the quantity of interest. Recently, several authors (Patil *et al.*, 1994; Muttlak and McDonald, 1990a, 1990b; Mode *et al.*, 1999) have suggested the use of RSS in the collection of environmental data. The efficiency of RSS was studied by Takahasi and Wakimoto (1968), Dell and Clutter (1972), Stokes (1995), and Barabesi and El-Shaarawi (2001).

RSS is a three-stage procedure. The first stage requires the selection of k^2 sampling units at random, which in turn are randomly partitioned into k groups of size k, where k is a design parameter to be chosen (usually a small number) by the investigator. In the second stage, for each set of k units, the units are ranked by some means, which does not involve actual measurement. At the final stage the unit with rank 1 is measured in the first set of k units, the unit with rank 2 is measured in the second set of k units, and this process of quantification continues until the unit with rank k is measured in the last set of k units. This process is eventually repeated for n cycles of k ordered sets in such a way that the final sample is of size nk. To illustrate the RSS selection procedure, consider a simple example of selection from the standard normal $N(0,1)$ population using R software and with $k = 3$ and $n = 1$. The steps are as follows:

1. A random sample of size $k^2 = 9$ was generated and this is taken to represent the population of fixed values, which in a real situation we do not know. This gives: $-0.3250, 0.4521, 0.0226, -2.5456, -1.0857, 0.2670, 0.5063, 0.1797, -0.8871$. Randomly partition the nine observations above into three groups of size $k = 3$ each, which gives: group 1 ($-1.0857, -0.3250, -2.5456$); group 2 ($0.1797, 0.0226, -0.8871$); group 3 ($0.5063, 0.2670, 0.4521$).

2. This is the step where we rank the observations within each group using some information other than the actual measurements. As carried out here, this corresponds to the case where the information used achieves a perfect ranking. This gives:

   Ordered G1: -2.5456 -1.0857 -0.3250
   Ordered G2: -0.8871 0.0226 0.1797
   Ordered G3: 0.2670 0.4521 0.5063

3. This is the actual measurement stage where only three measurements are made: the element ranked as the smallest in group 1 is measured, the element ranked as the second smallest in group 2 is measured, and finally the element ranked as the largest in group 3 is measured. These measurements give the ranked sample as: $-2.5456, 0.0226, 0.5063$. The average of these values produces the estimate of the population mean, which is. -0.6722.

After showing how to draw an RSS and use the measurements to estimate the population mean, the efficiency of the procedure is given. For this purpose, without loss of generality, it is sufficient to consider the case of a single cycle, i.e., $n = 1$. Let **X** be the matrix of order statistics, with $X_{(j,i)}$ giving the jth order statistic of the ith group ($i = 1, \ldots, k; j = 1, \ldots, k$). Then the RSS estimator of μ is the mean $\bar{X}_{RSS} = \sum_{j=1}^{k} X_{(j,j)}/k$ of the diagonal elements of **X**. Write $X_{(j,j)} = \mu + \sigma \varepsilon_{(j)}$, where

$\varepsilon_{(j)}$ is the jth order statistic of $\varepsilon_1, \ldots, \varepsilon_k$. Let $E(\varepsilon_{(i)}) = \alpha_i$ and $\text{Var}(\varepsilon_{(i)}) = \varpi_{ij}$. It can be shown that

$$E(\bar{X}_{\text{RSS}}) = \mu \quad \text{and} \quad \text{Var}(\bar{X}_{\text{RSS}}) = \frac{\sigma^2}{k^2} \sum_{i=1}^{k} \varpi_{ii} = \frac{\sigma^2}{k}\left(1 - \frac{\sum_{i=1}^{k} \alpha_i^2}{k}\right).$$

Therefore $\bar{X}_{\text{RSS}}$ is an unbiased estimator for μ and its relative precision (RP) compared to $\bar{X}$, the SRS estimator, is

$$RP = \frac{k}{k - \sum_{i=1}^{k} \alpha_i^2}.$$

Takahasi and Wakimoto (1968) have shown that $1 \leq RP \leq (n+1)/2$, with $RP = (n+1)/2$ when sampling from the uniform distribution. The relative efficiency of RSS to SRS, for $k = 2, \ldots, 10$ and for samples from a parent normal distribution, is given in Table 5.3. The relative efficiency was calculated using both the method of moments estimator and the maximum likelihood estimator of the population mean, and these are referred to as RPM and RPL, respectively. For the computation of RPM, the expected values of order statistics are required, which, for the normal distribution, are available in Pearson and Hartley (1976). It was shown by Barabesi and El-Shaarawi (2001) that RPL is well approximated as

$$RPL \approx 1 + 0.4805(k - 1).$$

Table 5.3 shows that RPL is always larger than RPM, as would be expected since the method of maximum likelihood is more efficient when the distributional form is known. However, the method of moments has the favorable property that it requires only the existence of the distributional mean and variance.

Table 5.3   Comparison of RPM and RPL for various values of k and for samples from a normal distribution.

k	*RPM*	*RPL*
2	1.470	1.961
3	1.914	2.442
4	2.347	2.922
5	2.770	3.403
6	3.186	3.883
7	3.595	4.364
8	3.999	4.844
9	4.399	5.325
10	4.794	5.805

The above results assume that the units within each random sample can be exactly ranked. Errors may arise, however, as a result of the ranker's judgment. In this case it has been shown (Dell and Clutter, 1972) that irrespective of ranking errors, $\bar{X}_{\text{RSS}}$ continues to be unbiased and is at least as efficient as $\bar{X}$ when based on the same number of measurements. To reduce the impact of ranking errors, McIntyre (1952) suggested replicating the RSS process several times or cycles. In each cycle, an RSS sample of size r is selected with r small enough to reduce the ranking errors.

5.5   Model-based sampling

5.5.1   A basic concept: estimation of population mean

In design-based sampling, the population values $Y_1, \ldots, Y_N$ are assumed to be known exactly and the only randomness involved is in the selection of the sample of size n. In model-based sampling, random selection of the sample is not required, instead $Y_1, \ldots, Y_N$ are assumed to be random variables, whose distribution is either fully or partially known. We refer the readers to Thompson (1997) for more details about this approach The fundamental difference between the two approaches can be illustrated by the following simple example.

Assume that $Y_1, \ldots, Y_N$ are independent random variables with a common normal distribution $N(\mu, \sigma^2)$ whose mean μ and variance σ^2 are unknown parameters. Since random selection is not involved, let the sample be $Y_1, \ldots, Y_n$.

$$\overline{Y} = \frac{1}{N} \sum_{i=1}^{N} Y_i = \frac{1}{N}(T_1 + T_2),$$

where $T_1 = \sum_{i=1}^{n} Y_i$ is known after sampling, while $T_2 = \sum_{i=n+1}^{N} Y_i$ is unknown. The problem is then reduced to estimating T_2 given the value of T_1. Under the model, T_1 and T_2 are independent $N(n\mu, n\sigma^2)$ and $N((N-n)\mu, (N-n)\sigma^2)$, respectively. Let λT_1 be a linear estimator of T_2, where λ is an unknown constant that is to be determined. Thus

$$Y^* = \frac{1+\lambda}{N} T_1,$$

and the error in this estimator is $m = \overline{Y} - Y^* = (T_2 - \lambda T_1)/N$. It is immediately clear that Y^* is an unbiased estimator of $\overline{Y}$ when $\lambda = (N-n)/n$. The mean square error of this estimator is

$$E(m^2) = \frac{N-n}{N} \frac{\sigma^2}{n}.$$

It should be noted the expression for the mean square error is the same as the one obtained for estimating the population mean based on SRS. But here, the result depends upon the assumption of independence and constancy of the mean and variance of Y, the variable being measured, over what would be the population of N elements. To take the inference further, a specific distributional form such as the normal distribution will generally be assumed.

5.5.2   Further comments about model-based sampling

The discussion introduced above can be generalized in at least two ways. The first case is where the mean and/or variance of the Y_i are dependent on explanatory variables. Interested readers can consult Hájek (1981), where some examples are provided. The second is the case where spatial and temporal correlations are present in addition to possibly dependence on explanatory variables. These are discussed in Cressie (1993) and Thompson (1997). Usually prediction at unknown locations or time is the aim of the analysis, and in this case kriging methods may be used in the design of the sampling network.

5.6   Concluding remarks

This chapter has discussed traditional sampling methods. These methods have application to sampling for the collection of microbiological data in the assessment of the quality of recreational waters. Simple random sampling is rarely used in this regard. The most common sampling techniques used are systematic sampling and stratified random sampling. Recently there has been interest in using composite and ranked set sampling, particularly when the resources available for field sampling are limited. We have discussed CS and RSS in some detail so as to provide microbiologists with information about the efficiency of these techniques. Finally, we have provided a preliminary concept about model-based sampling. The issue of using the combination of design-based and model-based sampling is beyond the scope of this chapter. It should be mentioned that Cochran (1946) compared the accuracy of SYS, STR and SRS using the concept of the superpopulation as given in the equations (5.4). Extension of Cochran's results to plane sampling is given by Quenouille (1949).

References

Aitchison, J., and Brown, J.A.C. (1981) *The Lognormal Distribution*. Cambridge University Press, Cambridge.

Barabesi, L., and El-Shaarawi, A.H. (2001). The efficiency of ranked set sampling for parameter estimation. *Statistics and Probability Letters*, 53: 189–199.

Barnett, V., and Bown, M. (2002) Statistically meaningful standards for contaminated sites using composite sampling. *Environmetrics*, 13: 1–13.

Bayley, G.V., and Hammersley, J.M. (1946) The 'effective' number of independent observations in an autocorrelated time series. *Supplement to the Journal of the Royal Statistical Society*, 8(2): 184–197.

Cochran, W.G. (1946) Relative accuracy of systematic and stratified random samples for a certain class of populations. *Annals of Mathematical Statistics*, 17: 164–177.

Cochran, W.G. (1977) *Sampling Techniques*, 3rd edition. John Wiley & Sons, Inc., New York.

Cramér, H. (1946) *Mathematical Methods of Statistics*. Princeton University Press, Princeton, NJ.

Cressie, N.A.C. (1993) *Statistics for Spatial Data*, revised edition. John Wiley & Sons, Inc., New York.

Dell, T.R., and Clutter, J.L. (1972) Ranked set sampling theory with order statistics background. *Biometrics*, 28: 545–553.

Dorfman, R. (1943) The detection of defective members of large populations. *Annals of Statistics*, 14: 436–440.

El-Shaarawi, A.H. (1989) Inference about the mean from censored water quality data. *Water Resources Research*, 25, 685–690.

El-Shaarawi, A.H. (2003a) Negative binomial distribution. In A.H. El-Shaarawi and W. Piegorsch (eds), *Encyclopedia of Environmetrics*, Vol. 3, pp. 1383–1388. John Wiley & Sons, Ltd., Chichester.

El-Shaarawi, A.H. (2003b) Lognormal distribution. In A.H. El-Shaarawi and W. Piegorsch (eds), *Encyclopedia of Environmetrics*, Vol. 2, pp. 1180–1183. John Wiley & Sons, Ltd., Chichester.

El-Shaarawi, A.H., and Damsleth, E. (1988) Parametric and nonparametric tests for dependent data. *Water Resources Bulletin*, 24(3), 513–519.

El-Shaarawi, A.H. and Marsalek, J. (1999) Guidelines for indicator bacteria in waters: uncertainties in applications. *Environmetrics*, 10: 521–529.

El-Shaarawi, A.H., and Lin, J. (2006) Interval estimation for log-normal mean with applications to water quality. *Environmetrics*, 18(1): 1–10.

El-Shaarawi, A.H., Esterby, S.R., and Dutka, B.J. (1981) Bacterial density in water determined by Poisson or negative binomial distributions. *Applied and Environmental Microbiology*, 41, 107–116.

Feller, W. (1968) *An Introduction to Probability Theory and Its Applications*, Volume 1, 3rd edition. John Wiley & Sons, Inc., New York.

Fisher, R.A., Thornton, H.G., and MacKenzie, W.A. (1922) The accuracy of the plating method of estimating the density of bacterial populations. *Annals of Applied Biology*, 9, 325–359.

Fisher, R.A. (1941) The negative binomial distribution. *Annals of Eugenics*, 11: 182–187.

Hájek, J. (1981) *Sampling from a Finite Population*. Marcel Dekker, New York.

Lehmann, E.L. (1983) *Theory of Point Estimation*. John Wiley & Sons, Inc., New York.

McIntyre, G.A. (1952) A method for unbiased selective samping using ranked sets. *Australian Journal of Agricultural Research*, 3: 385–390.

Mode, N.A., Conquest, L.L., and Marker, D.A. (1999) Ranked set sampling for ecological research: accounting for the total costs of sampling. *Environmetrics*, 10: 179–194.

Muttlak, H.A., and McDonald, L.L. (1990a) Ranked set sampling with respect to concomitant variables and with biased probability of selection. *Communications in Statistics: Theory and Methods*, 19(1): 205–219.

Muttlak, H.A., and McDonald, L.L. (1990b) Ranked set sampling with size biased probability of selection. *Biometrics*, 46, 435–445.

Patil, G.P., Sinha, A.K. and Taillie, C. (1994) Ranked set sampling. In G.P. Patil and C.R. Rao (eds) *Handbook of Statistics*, Vol. 12, pp. 167–200. North Holland, Amsterdam.

Pearson, E.S. and Hartley, H.O. (1976) *Biometrika Tables for Statisticians*, Volume II. Cambridge University Press, Cambridge.

Quenouille, M.H. (1949) Problems in plane sampling. *Annals of Mathematical Statistics*, 20, 353–360.

Stokes, S.L. (1995) Parametric ranked set sampling. *Annals of the Institute of Statistical Mathematics*, 47, 465–482.

Takahasi, K., and Wakimoto, K. (1968) On unbiased estimates of the population mean based on the sample stratified by means of ordering. *Annals of the Institute of Statistical Mathematics*, 20, 1–31.

Thompson, S.K. (1997) *Theory of Sample Surveys*. Chapman & Hall, New York.

US Environmental Protection Agency (1986) Ambient water quality criteria for bacteria – 1986, EPA440/5-84-002 Washington, DC.

Watson, M.A. (1936) Factors affecting the amount of infection obtained by Aphis transmission of the virus Hy. III. Philos. Trans. Roy. Soc. London, B, 226, 457–489.

Wymer, L.J., Brenner, K.P., Martison, J.W., Stutts, W.R., Schaub, S.A. and Dufour, A.P. (2005) The EMPACT Beaches Project: Results from a study on microbiological monitoring in recreational waters, US EPA 600/R-04/023, Washington, DC.

6

The lognormal distribution and use of the geometric mean and the arithmetic mean in recreational water quality measurement

Larry J. Wymer and Timothy J. Wade

National Exposure Research Laboratory, US Environmental Protection Agency, Cincinnati, OH, and National Health and Environmental Effects Research Laboratory, US Environmental Protection Agency, Research Triangle Park, NC

6.1 Introduction

Since 1968 United States recreational water quality criteria have set a limit on the geometric mean for fecal indicator bacteria from a number water samples taken over a period of time (National Technical Advisory Committee, 1968; US Environmental Protection Agency, 1976, 1986). On the other hand, for purposes of determining limits on effluents, including sewage, discharged into surface waters, the US EPA specifies that '[c]alculations for all limitations which require averaging of measurements shall utilize an arithmetic mean unless otherwise specified by the

Statistical Framework for Recreational Water Quality Criteria and Monitoring Edited by Larry J. Wymer
© 2007 John Wiley & Sons, Ltd

Director in the permit' (US EPA, 1980, 2003). These limits, a geometric mean criterion for beaches and arithmetic mean for discharges, both pertain to provisions of the Clean Water Act of 1977 as amended by the Beaches Environmental Assessment and Coastal Health (BEACH) Act of 2000.

In addition to this disagreement between types of means that are used in beach monitoring and those used in limiting discharges, a trio of papers published in the late 1990s evaluated uses of the geometric mean and reached conclusions such as 'the use of this statistic . . . is inappropriate for characterizing risk' (Haas, 1996), 'the arithmetic mean may generally provide a better approximation of the average risk' (Crump, 1998), and '[geometric means] should be phased out as regulatory criteria as soon as it is practical' (Parkhurst, 1998). Statements such as these serve to further create doubt about the appropriateness of geometric means in the minds federal and state regulators and stakeholders.

This chapter examines criticisms of the use of the geometric mean in risk assessment and environmental monitoring and evaluates its relevance to risk-based recreational water monitoring. Reasons for using the geometric mean (or rather the mean of the logarithms of the indicator densities, as we shall see) in modeling risk attributable to swimming in contaminated waters are explored and alternative models examined.

In the course of this discussion, we will refer to properties of the normal and lognormal probability distributions. For reference, a comparison of some characteristics of normal and lognormal distributions is presented in Table 6.1. The interested reader can find more detailed information and discussions in Crow and Shimizu (1988), Aitchison and Brown (1969), or Johnson and Kotz (1970), among other works. Estimation of lognormal parameters specifically in the context of environmental monitoring is discussed in Gilbert (1987).

Table 6.1   Characteristics of normal and lognormal distributions.

Distribution of x	Normal	Lognormal (2-parameter)
pdf of x	$\dfrac{1}{\sigma\sqrt{2\pi}}\exp\left[-\dfrac{(x-\mu)^2}{\sigma^2}\right]$ for $-\infty < x < \infty$, $\mu = E(x)$, $\sigma^2 = \mathrm{Var}(x)$	$\dfrac{1}{x\sigma_y\sqrt{2\pi}}\exp\left[-\dfrac{(\ln(x)-\mu_y)^2}{\sigma_y^2}\right]$ for $0 < x < \infty$, $y = \ln(x)$, $\mu_y = E(y)$, $\sigma_y^2 = \mathrm{Var}(y)$
Arithmetic mean (expected value) of x	μ	$\mu_x = \exp(\mu_y + \sigma_y^2/2)$
Geometric mean of x	—	$\exp(\mu_y)$
Median of x	μ	$\exp(\mu_y)$
Variance of x	σ^2	$\exp(2\mu_y + \sigma_y^2)\left[\exp(\sigma_y^2) - 1\right]$
Coefficient of variation of x	σ/μ	$\sqrt{\exp(\sigma_y^2) - 1}$

Source: Based on Gilbert (1987, Table 12.1)

6.2   Uses of the geometric mean in microbiology

Microbiological data frequently are re-expressed in terms of (usually base 10) logarithms, or 'log-transformed', prior to statistical analysis or summary. Such data include colony-forming units from plate counts, most probable number estimates, quantitative polymerase chain reaction cell equivalents, enzyme-linked immuno-sorbent assay counts, and unit volume densities derived from these quantities. Typically comparisons are subsequently made among the 'log means' of various groups, or 'log mean' microbial densities are used as either criterion or predictor variables in a model. Logically, the term 'log mean' would seem to indicate the logarithm of the mean, but it is commonly used to refer to the mean of the logarithms.

Although statistical analyses are typically done on the log-transformed values, the results are frequently re-evaluated in terms of their antilogs in order to express the results, such as means and their associated confidence intervals, in the same units as the original data. Taking the antilog of the log mean (again, mean of the logarithms) yields the geometric mean, defined as the nth root of the product of n sample values, or $GM=\exp(\mu_y)$ (when μ_y is expressed in natural logarithms, see Table 6.1). Hence, the geometric mean is often just a re-expression of the log mean. (This fact is nowhere made more evident than in the 1968 recommendation from the National Technical Advisory Committee that 'fecal coliform content of primary contact recreation waters shall not exceed a log mean of 200/100 ml'. Clearly, the meaning is *geometric mean*, not log mean.) Several reasons for using log-transformed data, leading to the log mean or geometric mean for bacteriological data, can be identified.

Firstly, the distribution of microbiological densities (number of organisms per unit volume) generally appears to be right-skewed. This tendency to exhibit a long right-side tail is shared by many diverse data series for which there exists a lower bound, usually zero, but no upper limit. Examples are household incomes, body weights, revenues of firms, and concentration of various pollutants in the air. Log-transformation of such data tends to result in a more nearly symmetric distribution, which may justify statistical testing and estimation procedures on the log-transformed data using the normal distribution (Table 6.1). For an argument why environmental sampling data in general should follow a lognormal distribution, see Esmen and Hammad (1977).

Secondly, using log-transformed data may result in a more appropriate statistical analysis. The standard deviation of microbiological data is commonly expressed as a coefficient of variation (CV), that is, the raw standard deviation divided by the mean. The motivation for this is the common observation that, while the absolute standard deviation increases with increasing mean, the CV appears to be more or less constant. Given a lognormal distribution, a constant CV implies a constant log variance (σ_y^2) since the lognormal CV depends solely upon σ_y^2 (see Table 6.1). Analyses of differences between two or more groups with respect to the means of their logarithms can utilize tests and estimation procedures based on the normal distribution with homogenous variance. Antilogs of these differences and their confidence limits have the natural and desirable (particularly in the field of microbiology) interpretation of ratios, or equivalently percentage differences between means. Given the homogeneity of variance (equality of the σ_y^2) these ratios will apply to either their arithmetic means or their medians (roughly, geometric mean), since these differ only in terms of σ_y as shown in Table 6.1.

Thirdly, a related point is relevant when a comparison of the CVs is desired. Such a comparison is usually of interest, for example, in the evaluation of a new method with respect to some standard method for enumeration of bacteria. In this case, an F-test of the ratio of their respective log variances $(\sigma_{y_1}^2 / \sigma_{y_2}^2)$ yields the comparison directly given that the lognormal CV depends solely upon σ_y^2.

Fourthly, depending upon the true variance and mean of the data, the variance of the arithmetic mean (or its square root, the standard error of estimation) may be larger than that of the geometric mean. Thus, the geometric mean is a more stable estimator than is the arithmetic mean under these conditions. This phenomenon will be discussed in greater detail later in the section on relative efficiency of the arithmetic mean and geometric mean estimators.

Finally, health models involving indicators commonly find some transformation of the exposure metric to be necessary for properly expressing the exposure–response relationship (or else they simply use exposure categories). Usually, this is the logarithm of indicator density (Cabelli, 1983; Dufour, 1984; Seyfried *et al.*, 1985; Ferley *et al.*, 1989; Wade *et al.*, 2006). In at least one case where another transformation had been promoted (a square root in Kay *et al.*, 1994), one subsequent reanalysis used the logarithmic form (van Dijk *et al.*, 1996) and a further review recommended that other forms for the exposure metric be investigated (Institute for Environment and Health, 2000). The logarithmic transformation has an advantage of being readily interpretable and explainable in regression models, in that the coefficient of a log-transformed predictor gives the amount of increase in the dependent variable that can be expected from a constant factor change (a tenfold change in the case of base 10 logarithms) in the predictor. Other transformations or polynomials do not lend themselves to such easy interpretation. To be consistent in the use of the models specifying log-transformed exposure, indicator density must be expressed as a log mean, which, as previously noted, 'back-transforms' to a geometric mean.

6.3   Criticism of the geometric mean as a measure of exposure or contamination

Several papers have criticized using the geometric mean as a measure of exposure or environmental contamination. Four major works are identified and summarized here.

Landwehr (1978) points out that the expected value of the geometric mean of the lognormal distribution depends upon sample size (n). In particular, it decreases with increasing sample size. For large n, the expected value equals the median of the lognormal distribution. Given this fact, a discharger could hope to achieve a smaller estimated geometric mean for some pollutant by simply taking more samples. Two plants that report the same geometric mean for a contaminant may have based their values on different sample sizes, in which case the two estimated geometric means will be more difficult to compare.

Figure 6.1 illustrates the behavior of the geometric mean as the sample size increases given a lognormal distribution of the data with a log variance that is typical for recreational water quality indicator densities (for example, see Table 4 in EPA, 1986). The general appearance and behavior of the sampling distribution is typical regardless of the location and shape (log variance) parameters of the lognormal distribution. In this case, with a true median of 60 and $\log_{10}$ standard deviation of 0.4 (consistent with Table 6.1, this corresponds to $\sigma_y \approx 0.92$), the expected value of the sampling distribution of the geometric mean goes from 92 to 67 to 61 as sample size increases from 1 to 4 to 32. This is a result of the decrease in the right tail area of the respective sampling distributions. The left side of the sampling distribution, particularly the mode, is seen to be *increasing* with sample size.

Haas (1996) finds that two mathematical models 'characterize all known pathogen oral dose-response relationships in humans', the exponential and beta-Poisson (see also Haas, 1983a, 1983b). If the pathogen density (number of

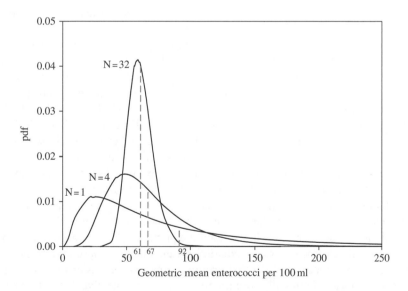

Figure 6.1 Sampling distribution and expected value of the geometric mean for $N = 1$, 4, and 32 from a lognormal distribution with a true median of 60 and $\log_{10}$ standard deviation of 0.4. The expected value of the geometric mean is seen to converge to the value of the median rather rapidly in this case.

organisms per unit volume) is m, and the volume of water (or other medium) ingested is V, the exponential model gives the probability of infection as

$$p = 1 - \exp\left(-\frac{mV}{k}\right). \tag{6.1}$$

The single parameter, k, represents the average number of organisms that must be ingested to cause an infection. The beta-Poisson model gives the probability of infection as

$$p = 1 - \left(1 + \frac{mV}{\beta}\right)^{-\alpha}. \tag{6.2}$$

Here, β corresponds to the median infectious dose, and α is a 'slope' parameter (this basically determines the steepness of the dose–response curve). The beta-Poisson model approaches the exponential in the limit as $\alpha \to 1$. At small doses (which one hopes to be the case), the beta-Poisson relationship becomes approximately linear, in which case

$$p \approx \frac{\alpha V}{\beta} m. \tag{6.3}$$

Haas shows that in either case, the exponential or the linearized beta-Poisson, use of the arithmetic average for the dose yields the correct estimate for the expected proportion of illness among the exposed population.

Crump (1998) is concerned that often only a group summary exposure level is available in a risk analysis and thus the risk assessment must proceed as if everyone in that group were exposed at the same level. In this case, his criterion for deciding if the arithmetic or geometric mean is the better exposure summary to use is which one gives a result closer to the actual risk that would be found if each individual's exposure level were known. Using this criterion, he finds that the arithmetic mean is preferred over the geometric mean whenever the dose–response function is convex (i.e., has increasing slope) in the region of interest. For cases when the dose–response function is not convex, he examined several reasonable data sets and determined the arithmetic mean still to be the preferred summary measure.

Parkhurst (1998) maintains that geometric means 'should be abandoned in favor of arithmetic means, unless they are clearly shown to be preferable for specific applications'. This statement is based upon the premises (1) that the geometric mean is a biased estimator of the true population mean, which is appropriately estimated by the arithmetic mean, and (2) that arithmetic means are easier to calculate and understand, more scientifically meaningful, and more protective of public health.

The first point is clear. The arithmetic mean is an unbiased estimator of the true population mean regardless of the underlying probability density function (pdf – the only requirement being that the pdf possesses a finite first and second moment). However, the geometric mean is not an unbiased estimator of the true population mean regardless of the underlying pdf. If the pdf is lognormal, the geometric mean is a biased estimator of the population mean, but a consistent (i.e., asymptotically unbiased) estimator of the population median. Also, if the data have a lognormal distribution, the arithmetic mean is not an efficient estimator of the population mean. That is, there is an estimator (see Gilbert, 1987, p. 165) that possesses a lower variance than that of the arithmetic mean. Parkhurst discusses this, referring to the efficient estimator as the 'corrected geometric mean'.

In saying that arithmetic means are 'more scientifically meaningful', Parkhurst is referring to their usefulness in mass balance equations and fate and transport models. If one knows the arithmetic mean of some contaminant in each of two streams which join together to form a single stream, the resulting concentration in the single stream can be readily calculated. However, the same is not true for the geometric mean. Concentration data are commonly used in determining the fate and transport of contaminants, and representing the concentration data by geometric means will result in underestimating the total amount of that contaminant a community is exposed to over a fixed period of time. Using the arithmetic mean in such calculations yields the correct exposure.

6.4 Counterpoints to criticisms of the geometric mean as a measure of exposure or contamination

Landwehr's argument is undeniably true as for as the expected value of the geometric mean is concerned. The implication is that if geometric mean sample results for pollution levels were publicly posted, then a discharger could look better over the long term by taking more samples. This is true whether the plant is a relatively clean discharger or a polluter. This would be an important consideration with respect to environmental regulations if regulations were to involve, say, a ranking of dischargers. Beach managers could take advantage of this quirk in the geometric mean if they relied on postings of an average fecal indicator level at the beach.

However, environmental regulations, and specifically the sort of recreational water standards we are concerned with here, generally establish a fixed limit for the contaminant or indicator. The fate of a polluter is much different under the case in which it must not exceed this limit, regardless of how its mean discharge level compares to others. Figure 6.2 illustrates the effect of increased sample size on the likelihood that a discharger's geometric mean will exceed a limit of 33 (which happens to be the limit for the one-month geometric mean enterococci per 100 ml at freshwater beaches in the United States) when the true median (see Table 6.1) is 60 and its $\log_{10}$ standard deviation is 0.4. One can see that, although the expected value of the geometric mean does indeed decline, the probability of exceedance (indicated by the proportion of the curve that is within the shaded area) goes up until, at a sample size of 32, the polluter is virtually assured that it will be found in violation. This illustrates the general fact that, paradoxically, whenever its median amount of

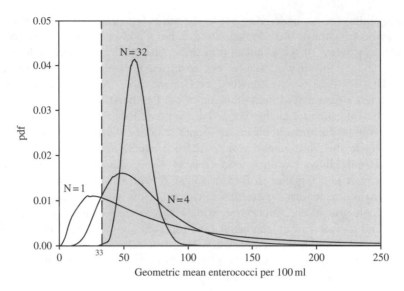

Figure 6.2   The scenario is the same as in Figure 6.1, $N = 1$, 4, and 32 from a lognormal distribution with a true median of 60 and $\log_{10}$ standard deviation of 0.4. The probability of exceeding a geometric mean compliance limit of 33 enterococci per 100 ml (dashed line) is equal to the shaded area under the pdf curve of the sampling distribution. While, as Figure 6.1 illustrates, the expected value of the geometric mean does indeed decline for increasing N, the likelihood that the sample geometric mean will exceed the compliance limit increases, consistent with desirable sampling properties.

discharge is truly higher than some geometric mean regulatory limit, a polluter's strategy of lowering the expected value of its geometric mean by increasing sample size will actually increase its probability of being out of compliance.

Haas and Crump are concerned with the use of what epidemiologists refer to as *ecologic data*, that is, summary measures of central tendency, such as geometric or arithmetic means for a specific population, to estimate the expected incidence of a disease when an exposure–response relationship based on *individual*-specific exposure to a *pathogen* is known. However, exposure–response relationships that pertain to recreational waters are (1) based on indicators, not pathogens, and (2) based on exposure measurements that are only indirectly related to individual exposures or dosages. The beta-Poisson or exponential dose–response model may indeed be correct with respect to any single pathogen that is responsible for swimmers' illnesses. Indicator organisms, however, are at least two steps removed from these pathogens–indicators indicate (imperfectly) the degree of fecal pollution, which in turn suggests (imperfectly) the presence of pathogens. Add to this the further complications that various kinds of pathogens (possibly including unknowns) may be present in various numbers and that their virulence may vary depending on their condition (e.g., stressed or not).

Most studies on swimmers' health, conducted both in the USA (Cabelli, 1983; Dufour, 1984) and elsewhere (Cabelli, 1983; Cheung *et al.*, 1990; Corbett *et al.*, 1993; Dufour, 1984; Medema *et al.*, 1995; Prieto *et al.*, 2001), have utilized ecologic exposure data, specifically the log mean, not the individual exposure for each swimmer (studies in which individual data on illness as well as other characteristics, such as age or gender, are used in conjunction with ecologic exposure data are often referred to as 'semi-individual' studies). The exposure measurement is, therefore, precisely the measurement that one would obtain in practice, that is, an average value. Even studies which purport to measure individual-specific exposure do not obtain very precise estimates of exposure, at best defining exposure as the concentration of *indictor bacteria* in a *single sample* collected within *10 minutes* and *10 meters* of the location where the individual may have actually been exposed (Kay *et al.*, 1994). Therefore, the quandary that is addressed by Haas and Crump – how to handle ecologic data on exposure when a personal exposure relationship is known – is not an issue in modeling bathers' illnesses.

Parkhurst brings up several valid concerns about using a geometric mean. First, it is true that the geometric mean is a biased estimate of the population mean for a lognormal – or indeed any skewed – distribution. We do agree that one should not use a geometric mean on those occasions when an estimate of the population average or total is required, such as in mass balance equations. In this context, Parkhurst does allow for the use of the geometric mean on one condition, stating, 'Geometric means might reasonably be used only in situations where means of consistently small samples are needed for data drawn from log-normal distributions'. We note that, with regard to monitoring beach water quality, whatever summary measure is developed, be it an arithmetic mean, a geometric mean, or some other form, is not intended to be used as input to a mass balance equation or the like.

Parkhurst's assertion that the arithmetic mean is more protective of public health, at least in this case, hinges on his point that it is 'more scientifically meaningful'. In a trivial sense an arithmetic mean is more protective of public health because it is always higher than the geometric mean – requiring the arithmetic mean rather than the geometric mean to meet a given numerical standard will result in a lower health risk. However, we would assume that any rule, whether based on a geometric or an arithmetic mean, is exactly as protective of public health in terms of tolerable health risks as deemed desirable, no more and no less. Otherwise, if, say, a lower risk were desirable – that is, the cost (*e.g.*, loss of recreation) of lowering the risk is outweighed by the benefit of doing so – the rule is flawed. It is only in the sense that an arithmetic or geometric mean may enable better *evaluation* of the health risks that one or the other may legitimately be said to be more protective of public health. The question becomes, then, which is the better *predictor* of health effects, the arithmetic mean or the geometric mean indicator density? Or perhaps it is some altogether different statistic.

Among Parkhurst's four points concerning the superiority of the arithmetic over the geometric mean, then, the one that holds sway is that the arithmetic mean is 'more scientifically meaningful', or, in our case, a better predictor of health effects. Parkhurst focuses on the usefulness of arithmetic means for estimating total contamination, and with this we are in complete agreement. In the United States, the Clean Water Act 1977 authorizes the US EPA to set limits on effluents discharged into surface waters, including those used for recreation, establishing the National Pollutant Discharge Elimination System (NPDES) for this purpose. Since the inception of the NPDES program, the EPA has specified that 'Calculations for all limitations which require averaging of measurements shall utilize an *arithmetic* mean unless otherwise specified by the Director in the permit' (US EPA, 1980, 2003; emphasis added). This condition applies to all NPDES permits, including those administered under state programs. However, the kind of recreational water quality monitoring that we address here is performed for the purpose of determining how 'safe' the water is for swimming, that is, how likely it is that swimmers will become ill from swimming in these waters. Therefore, the criterion for whether one measure or another is more scientifically meaningful in our case must be which one is a better predictor of health effects on those exposed to contamination in recreational waters. The next section discusses this issue.

6.5   Predicting health effects of swimming

As we stated at the beginning of this chapter, expressing microbial data in general and recreational water indicator densities in particular by using geometric means is often just an offshoot of using mean log values (mean of the logarithms). The US EPA conducted studies on swimmers' health in the late 1970s and early 1980s (Cabelli, 1983; Dufour, 1984) and derived exposure–response models for freshwater and marine beaches based on mean $\log_{10}$ densities per 100 ml of *E. coli* and enterococci in the water, these organisms being used as indicators of human fecal contamination. In order to use these models, then, it is necessary to find the mean of the log-transformed indicator counts per 100 ml from a sample. For example, the freshwater model for 'highly credible gastrointestinal illness' (HCGI) indicates that when mean $\log_{10}$ *E. coli* per 100 mL is equal to 2.1, about 9 extra cases of HCGI per 1000 swimmers can be expected. The 9 extra cases of HCGI per 1000 swimmers (or risk equal to 0.009) represents *attributable risk*, that is, the incremental risk that may be ascribed directly to swimming in contaminated waters. The antilog of 2.1 is 126. Using a target attributable risk of 0.009, criteria for freshwater beaches that were subsequently published by the US EPA (1986) and based on this health model specify a geometric mean of 'not less than five samples equally spaced over a 30-day period' not to exceed 126 *E. coli* per 100 ml.

From this, in order to satisfy the requirements of the health model, in particular the requirement that the mean of log indicator densities be calculated, the criteria for recreational water quality are necessarily expressed in terms of mean log densities, or, equivalently, the geometric mean. Note that $\log_{10}(GM) =$ mean $\log_{10}$(indicator density), where GM is the geometric mean. There is no corresponding model that involves the arithmetic mean, nor did the authors indicate any results of an attempt to fit the health effects model to the arithmetic mean, or to the log of the arithmetic mean. Therefore, in a sense, the geometric mean is the more protective of public health by default, simply because nothing is known of the relationship between health and the arithmetic mean.

It is not certain whether the fact that a roughly linear relationship exists between swimmers' illnesses and the logarithm of fecal indicator densities is a point in favor of the geometric mean. After all, the relationship could possibly be just as well, or better, described using the logarithm of the arithmetic mean rather the mean of the logarithms (which equals the logarithm of the geometric mean). It is interesting that this sort of linear relationship has been shown by the EPA data. What is more, the relationship obtained by the EPA has been confirmed most recently in work done by Kay *et al.* (2004) using the aforementioned UK data in order to interpret these data, which are based on more or less individual-specific exposures, when only a mean for a lognormal distribution of individual exposures is known. The resulting UK 'ecologic exposure' model is compared to the existing EPA model for enterococcus in marine waters in Figure 6.3. Relative risk of swimmers to non-swimmers as

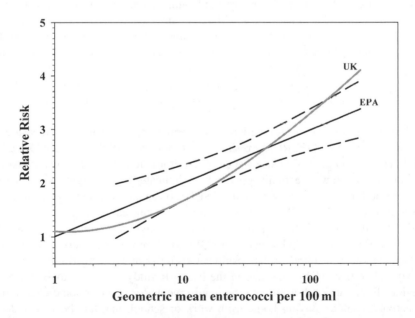

Figure 6.3 Predicted relative risk of gastroenteritis from UK ecologic risk (Kay *et al.*, 2004), and EPA HCGI (Cabelli, 1983) models. Relative risk is calculated as the ratio of total swimmer risk to background risk, where background risk is from non-swimmer illness rates. 95 % confidence limits for the EPA risk model are shown over the approximate range of the study data.

predicted by the two models, rather than attributable risk (absolute risk difference), is shown because of differences in the definition of gastrointestinal illness used in these studies. Some curvature in their resulting model is evident. However, the curvature is far less than it would be if risk were proportional to raw indicator levels, and the curve is generally within the 95 % confidence interval for the EPA model (the corresponding error for the UK model is not known, but could only lead to the conclusion that there is even less difference between these models). It is important also to note that the UK curve is based on a *known* geometric mean, while the EPA model has been developed based on *sample* geometric means. Were the UK model to be amended to account for sampling error, the resulting curve would 'flatten out' somewhat and appear even more similar to the EPA curve.

More recently, additional recreational water health studies have been conducted by the US EPA (Wade *et al.*, 2006). The primary purpose of the National Epidemiological and Environmental Assessment of Recreational (NEEAR) Water Study was to examine and model health effects when indicators of contamination are used other than plate counts from membrane filtration assays of *E. coli* and enterococci. In particular, a 'rapid method', quantitative polymerase chain reaction (QPCR), for detecting the presence of these organisms in water was investigated. QPCR detects and quantifies DNA from organisms, and while membrane filtration will detect only viable, culturable cells, QPCR will detect any cells that contribute to the DNA pool, living or dead. As of this writing, studies at four Great Lakes beaches had been completed. Data from these studies can be used for our purposes to evaluate relationships between swimmers' illnesses and compare the arithmetic mean, the geometric mean, the mean of the logarithms, and logarithm of the mean of QPCR 'cell equivalents' (QPCRCE), or, indeed, any other function of QPCRCE that we may desire.

Frequency histograms for various transformations and summary measures of the daily average of QPCRCE are shown in Figure 6.4. While all transformations exhibited some degree of positive skew, the arithmetic mean was severely skewed (the degree of skew is actually greater than indicated in the figure since several extreme outlying measures were excluded for presentation), and the mean of the log-transformed QPCRCE appeared the most symmetrical. Based on the NEEAR study, Wade *et al.* (2006) presented a model for swimming-related risk of acute gastrointestinal illness (AGI) relative to QPCRCE, reproduced here in Figure 6.5, and again indicating a linear relationship between risk and the logarithm of indicator density. Note that AGI as defined in the NEEAR study is similar to HCGI of the previous EPA studies, but without the requirements for fever or some clear impact on activity, such as staying home from work or school, that had been included in the definition for HCGI.

Results of comparison between different QPCRCE summary measures and swimmers' health are shown in Table 6.2. Here, logistic models were estimated for the log odds of swimmers' AGI as a function of the respective indicator measurements at 8:00 a.m. on the day of exposure, and dummy variables for

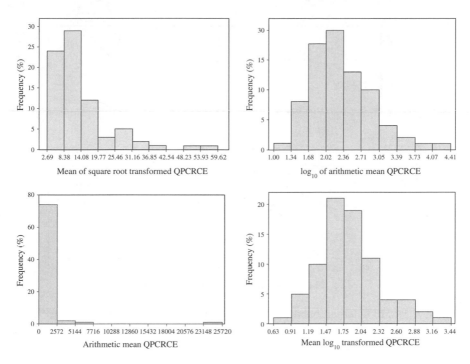

Figure 6.4   Frequency histograms for different transformations of daily averages of enterococcus QPCRCE. Based on 1482 measures from four Great Lakes beaches over 78 study days.

beach and age of swimmer. The requirement for exposure (i.e., the definition of a swimmer) is that the individual had full body contact in the water. The various means were calculated from QPCRCE per 100 ml results from six locations in the water at 8:00 a.m. Table 6.2 gives the estimated odds ratio per quartile for the respective indicator average. By far the strongest, and in fact the only statistically significant, association is evident for the mean $\log_{10}$ density, or equivalently $\log_{10}$ of the geometric mean. The lack of an observable relationship between the arithmetic mean, the geometric mean, or some other transformations (not shown) and AGI could be a result of the strong positive skewness of these measures.

Transformation to a log scale may result in a linear relationship between the QPCRCE and risk of illness, making effects more apparent in regression models. However, note that even the logarithm of the arithmetic mean fails to be a good predictor of swimmers' illnesses. This allows one to say that the geometric mean is more protective of public health than is the arithmetic mean in this instance, by virtue of their link with AGI risk through a logarithmic transformation. We will discuss the minimum variance unbiased estimator for the logarithm of the population mean (MVUE in Table 6.2) later. First, to provide some background, we examine

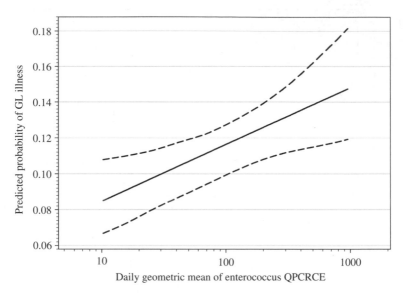

Figure 6.5   Predicted probabilities of gastrointestinal illness as a function of enterococcus QPCRCE, predicted from the logistic regression model, adjusted for age and beach (from Wade *et al.*, 2006).

Table 6.2   Results of logistic regression of various exposure measures (at 8:00 a.m.) on the probability of swimmers' gastrointestinal illness, adjusted for beach and age effects.

Measure of exposure[a]	Estimated odds ratio (OR)[b]	Standard error	Z^c	$P[H_0:OR=1]^c$	95 % Confidence interval for OR	
					Lower limit	Upper limit
$\log_{10}(GM)$ [=mean $\log_{10}(c)$]	1.38	0.13	3.302	0.001	1.14	1.66
$\log_{10}(AM)$	0.99	0.10	−0.105	0.916	0.82	1.20
MVUE $\log_{10}(\mu_x)^d$	0.91	0.05	−1.583	0.113	0.81	1.02
GM	1.00	0.00	1.387	0.166	1.00	1.01
AM	1.00	0.00	−1.883	0.060	1.00	1.00

[a] GM = geometric mean, AM = arithmetic mean, c = QPCRCE (QPCR cell equivalents)
[b] Odds ratio for an increase in the exposure metric from one quartile to the next
[c] Z, P-value for the null hypothesis that the odds ratio (column 1) is 1 (i.e., no effect)
[d] Minimum variance unbiased estimate of $\log_{10}(\mu_x) = \log_{10}$ (e) $\cdot (\bar{y} + s_y^2/2)$, where $y = \ln(x)$.

the raw, untransformed geometric and arithmetic mean and their efficiencies as estimators of central tendency.

6.6 Relative efficiency of the arithmetic mean and geometric mean estimators

Parkhurst and Stern (1998) discuss the relative efficiency of the geometric mean as an estimator of the arithmetic mean as a reason why it may be preferable to use the geometric mean when an estimate of the arithmetic mean is required. A sample arithmetic mean is an unbiased estimator of the mean for lognormally distributed data (or for data from any probability distribution). The geometric mean is a biased, but consistent, estimator of the median of a lognormal distribution, but underestimates the lognormal arithmetic mean. If one wishes to compare the performance of these two as estimators of the lognormal mean, it will be necessary then to compare the mean square error (MSE) of the geometric mean with the variance of the arithmetic mean, since the MSE accounts for both bias and variance of the geometric mean (Parkhurst and Stern, 1998). Table 6.3 gives the ratio of the variance of the arithmetic mean to the MSE of the geometric mean for a range of sample sizes and variances on the natural log scale (El-Shaarawi, personal communication, 2004).

For a given sample size the efficiency of the geometric mean relative to the arithmetic mean (the ratio in Table 6.3) increases as the variance increases and decreases as the sample size increases. Greater efficiency of the geometric mean is indicated by any combination of sample size and variance for which the ratio is larger than unity. The relative efficiency, R, is plotted against the standard deviation of the base 10 logarithms in Figure 6.6. Base 10 logarithms are commonly used in microbiological contexts, but the equation for R in Table 6.3 is simpler in terms of natural logarithms. Standard deviations can be converted to base 10 logarithms simply by multiplying the natural log standard deviation by a factor of 0.434 [$=\log_{10}(e)$]. US EPA (1986) criteria for fresh and marine recreational waters cite 'default' $\log_{10}$ standard deviations of 0.4 and 0.7, respectively, among weekly samples and specify that geometric means be based on at least five samples. Wymer *et al.* (2005) found spatial $\log_{10}$ standard deviations of 0.3 to 0.5 (along the same water depth) and temporal $\log_{10}$ standard deviations (day to day) ranging from 0.4 to 0.7 among five freshwater and marine beaches.

It is apparent that the ratio exceeds unity and, thus, the geometric mean is a more efficient estimator of the lognormal mean in almost all cases considered here when the sample size is below 20. The implication is that, although the geometric mean is a biased estimator of the lognormal mean, that is, 'over the long run' the average of all the sample geometric means will underestimate the lognormal mean, the geometric mean, in fact, will tend to be closer in value to the true mean than will the arithmetic mean for sample sizes and values of log standard deviation typically found in beach monitoring.

Table 6.3   Relative efficiency[a] of the geometric mean compared to the arithmetic mean as a function of natural log variance (ln Var) and sample size.

ln Var (σ_y^2)	Sample size (n)									
	1	3	5	7	9	11	13	15	17	19
0.1	1.00	1.07	1.04	1.00	0.96	0.93	0.89	0.86	0.83	0.80
0.2	1.00	1.14	1.09	1.01	0.94	0.88	0.82	0.77	0.73	0.69
0.3	1.00	1.23	1.14	1.03	0.94	0.85	0.78	0.72	0.67	0.62
0.4	1.00	1.31	1.19	1.06	0.94	0.84	0.76	0.69	0.63	0.58
0.5	1.00	1.41	1.26	1.08	0.94	0.83	0.74	0.67	0.61	0.56
0.6	1.00	1.51	1.32	1.12	0.96	0.84	0.74	0.66	0.60	0.55
0.7	1.00	1.62	1.39	1.16	0.98	0.85	0.74	0.66	0.59	0.54
0.8	1.00	1.75	1.47	1.21	1.01	0.86	0.75	0.66	0.60	0.54
0.9	1.00	1.88	1.56	1.26	1.04	0.88	0.76	0.67	0.60	0.54
1.0	1.00	2.02	1.65	1.32	1.08	0.91	0.78	0.69	0.61	0.55
1.1	1.00	2.18	1.75	1.38	1.12	0.94	0.81	0.71	0.63	0.57
1.2	1.00	2.35	1.86	1.45	1.17	0.98	0.84	0.73	0.65	0.58
1.3	1.00	2.53	1.98	1.53	1.23	1.02	0.87	0.76	0.67	0.61
1.4	1.00	2.73	2.11	1.61	1.29	1.07	0.91	0.79	0.70	0.63
1.5	1.00	2.95	2.25	1.71	1.36	1.12	0.96	0.83	0.73	0.66
1.6	1.00	3.19	2.41	1.81	1.44	1.18	1.01	0.87	0.77	0.69
1.7	1.00	3.45	2.58	1.93	1.52	1.25	1.06	0.92	0.81	0.73
1.8	1.00	3.73	2.76	2.05	1.61	1.32	1.12	0.97	0.86	0.77
1.9	1.00	4.04	2.96	2.19	1.71	1.40	1.19	1.03	0.91	0.81
2.0	1.00	4.38	3.17	2.33	1.83	1.49	1.26	1.09	0.96	0.86
2.1	1.00	4.75	3.41	2.50	1.95	1.59	1.34	1.16	1.02	0.91
2.2	1.00	5.15	3.67	2.67	2.08	1.70	1.43	1.24	1.09	0.97
2.3	1.00	5.59	3.95	2.87	2.23	1.82	1.53	1.32	1.16	1.04
2.4	1.00	6.07	4.25	3.08	2.39	1.94	1.64	1.41	1.24	1.11
2.5	1.00	6.59	4.59	3.31	2.56	2.09	1.76	1.52	1.33	1.19
2.6	1.00	7.17	4.95	3.56	2.76	2.24	1.89	1.63	1.43	1.28
2.7	1.00	7.80	5.35	3.84	2.97	2.41	2.03	1.75	1.54	1.37
2.8	1.00	8.48	5.79	4.14	3.20	2.60	2.18	1.88	1.65	1.48
2.9	1.00	9.24	6.26	4.47	3.45	2.80	2.35	2.03	1.78	1.59
3.0	1.00	10.06	6.78	4.83	3.72	3.02	2.54	2.19	1.92	1.72

Source: A. El-Shaarawi, personal communication, 2004
[a] The relative efficiency is given by:

$$R = \frac{\exp(2\sigma_y^2) - \exp(\sigma_y^2)}{n\left[\exp(2\sigma_y^2/n) - 2\exp\left((n+1)\sigma_y^2/2n\right) + \exp(\sigma_y^2)\right]}.$$

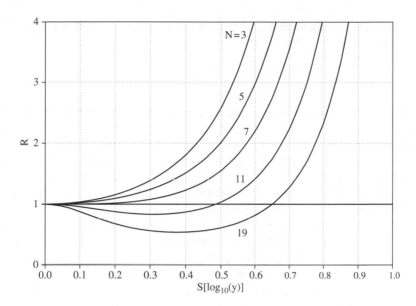

Figure 6.6   Relative efficiency (R) of the arithmetic mean to the geometric mean from lognormal populations as a function of $\log_{10}$ standard deviation (S) and sample size (N). R is defined as the ratio of the mean square error of the arithmetic mean to that of the geometric mean. Because the arithmetic mean is not a biased estimator of the expectation, its mean square error is equal to its variance.

6.7   Characteristics and relative efficiency of estimates on the logarithmic scale

The preceding shows that the geometric mean is likely to have a lower MSE than the arithmetic mean with respect to the true mean, given the usual range of sample sizes and standard deviations of log-transformed data relevant to recreational water sampling. However, we again emphasize that risk models for swimmers' illnesses generally have not used either type of mean directly, instead finding the best relationships between swimmers' health and logarithm of indicator densities. Therefore, it is appropriate to consider the sampling properties of logarithmic analogs to the geometric and arithmetic means, the mean of the logarithms and the logarithm of the (arithmetic) mean.

   The variance of the mean of the logarithms (here is where the customary term 'log mean' may be confusing, since we are also considering the logarithm of the arithmetic mean) is σ_y^2/n, where σ_y^2 is the variance of the log-transformed values and n is the sample size. From Table 6.1, it may be noted that the (natural) logarithm of the true mean for the lognormal distribution is $\mu_y + \sigma_y^2/2$, which we denote as θ.

The uniform minimum variance unbiased estimator (MVUE) of θ is $\hat{\theta} = \bar{y} + s^2/2$, and its variance is

$$\text{Var}(\hat{\theta}) = \frac{\sigma_y^2}{n} + \frac{0.5\sigma_y^4}{n-1} \tag{6.4}$$

(Lehman, 1983), where $\bar{y}$ is the sample mean of the log-transformed values and s^2 their sample variance. Here, the situation is reversed from that of estimation for the arithmetic mean, discussed previously, where estimators derived from logarithmically transformed values yield biased estimators of the arithmetic mean. The logarithm of the arithmetic mean is a biased estimator of the logarithm of the true lognormal (arithmetic) mean, and the best unbiased estimator is obtained in terms of the mean and variance of the logarithms.

By way of analogy to the use of a geometric mean as an estimator for the arithmetic mean, which was discussed above, consider the use of the mean of the logarithms, $\bar{y}$, as an estimator of the logarithm of the mean (which is best estimated by $\hat{\theta}$). As an estimator of θ, $\bar{y}$ is biased by an amount $-0.5\sigma_y^2$ so that its MSE is $\sigma_y^2/n + 0.25\sigma_y^4$. Thus, the relative efficiency of $\bar{y}$ relative to $\hat{\theta}$ (equation (6.4) divided by the MSE of $\bar{y}$), is greater than 1 only for sample size $n = 2$ ($\hat{\theta}$ does not exist for $n = 1$). This shows that the argument that the geometric mean can be a better estimator for the arithmetic mean for small sample sizes does not carry over into the realm of log-transformed values.

One may in practice use the logarithm of the arithmetic mean in place of $\hat{\theta}$, even though this estimator is not unbiased and the unbiased estimator, $\hat{\theta}$, is easily calculated. An analytical expression for the relative efficiency of the logarithm of the arithmetic mean with respect to the uniform MVUE is not available. We could use Monte Carlo simulation to find the relative efficiency over a range of parameter values. However, for our purposes, we simply note that neither the logarithm of the arithmetic mean QPCRCE nor the MVUE, $\hat{\theta}$, showed a significant relationship to AGI in the NEEAR Study (see Table 6.2).

6.8 Conclusions

The papers cited above raise valid concerns about the use of geometric means and present powerful arguments in favor of using arithmetic means as summary measures under specific circumstances. These concerns have been noted by those responsible for establishing criteria for recreational water quality and by those responsible for monitoring water quality in accordance with these criteria.

Haas's and Crump's objections to using a geometric mean are not particularly relevant to the relationship between fecal indicator bacteria and swimmers' health. They are mentioned here because of concerns that their papers have raised among recreation water quality policy makers and managers. Their research is based on the traditional dose–response relationship, that is, a risk model based on exact, individual-specific dosages. Attempts at relating swimming-related risk to a somewhat individual-specific measure of exposure have been made (Kay *et al.*, 1994). However, even these are not traditional dose–response models since they still rely only on measures of *potential* exposure (i.e., potential ingestion of sewage-contaminated water), not on measures of *dosage* (i.e., ingestion of specific pathogens). Most exposure–response relationships for swimming-related risks of illness have been developed specifically for the practical case in which only a summary measure of water quality for a beach as a whole is available. At best, the summary measurement, whatever its form, provided by any recreational water monitoring program provides an inexact evaluation of sewage pollution at the beach, which in turn only suggests the likely prevalence of pathogens in the water, which may comprise any number of different organisms.

Landwehr's paper illustrates an interesting paradox. Given that a polluter truly exceeds some standard that is expressed in terms of a geometric mean, by increasing the number of samples collected it can decrease the expected value of its geometric mean, but only at the expense of increasing the likelihood that it will be found to be in violation. The paradox is resolved when one considers the behavior of the sample mean of the concentration logarithms. Regardless of the fact that the expected geometric mean decreases with increasing sample size, the behavior of the geometric mean is always consistent with respect to the probability of rejecting a beach as being 'safe'. If such rejection is warranted according to its 'true' geometric mean, the probability of rejection increases with increased sample size, and, otherwise, decreases with increased sample size.

Parkhurst's comments on the use of geometric means with respect to limits on allowable discharges are well founded. Geometric means of concentrations are not useful for calculating loadings. Additionally, a geometric mean can be subject to manipulation. There exists the opportunity for a discharger to adjust the amounts of pollutants that are discharged, increasing the variance of emissions so that the geometric mean is decreased without any change in the total amount of pollutants that are released into the environment. This applies mainly to a temporal geometric mean, since discharge limits are generally applied 'at the pipe'. From Table 6.1, when the arithmetic mean, μ_x, is held constant, an increase in σ_y^2 (the 'log variance') results in lower $\exp(\mu_y)$, the geometric mean, given $\mu_x = \exp(\mu_y + \sigma_y^2/2)$. However, a manipulation of this nature would not appear to be a problem in beach monitoring, nor are beach monitoring, data intended to be used in mass balance calculations.

Analyses of swimmers' health effects from the studies on QPCR indicate that the mean of the logarithm of QPCRCE is strongly associated with the incidence of GI, while other summary measures of QPCRCE data fail to show any relationship. Mean log QPCRCE is more protective of public health than is the arithmetic mean QPCRCE, or, as also has been shown, even the logarithm of the arithmetic mean, based on the definition of 'more protective of public health' being 'a better predictor of health effects'. Thus we conclude that recreational water quality criteria are correctly based on the mean of the log QPCRCE from among the samples collected at the beach. Because of the direct correspondence between mean log and geometric mean, the criteria may be translated into geometric means with no loss in meaning.

As for recreational water quality criteria based on membrane filtration (plate counts) from marine or freshwater beaches, these must continue to be expressed in terms of geometric means (actually, mean logs) by default. In the USA, the only models for human health effects that are available from studies are those that relate attributable risk to mean log indicator density. If arithmetic mean indicator density were available, we simply do not have a way to deal with it! Note that a similar argument for using a geometric mean when the exposure–response model is log-linear has been used with respect to occupational exposures (Seixas *et al.*, 1988).

References

Aitchison, J., and Brown, J.A. (1969) *The Lognormal Distribution*. Cambridge University Press, Cambridge.

Cabelli, V.J. (1983) Health effects criteria for marine recreational waters, EPA-600/1-80-031. US Environmental Protection Agency, Research Triangle Park, NC.

Cheung, W.H., Chang, K.C.K., Hung, R.P.S., and Kleevens, J.W.L. (1990) Health effects of beach water pollution in Hong Kong. *Epidemiology and Infection*, 105(1): 139–162.

Corbett, S.J., Rubin, G.L., Curry, G.K., and Kleinbaum, D.G. (1993) The health effects of swimming at Sydney beaches. The Sydney Beach Users Study Advisory Group. *American Journal of Public Health*, 83(12): 1701–1706.

Crow, E.L., and Shimizu, K. (1988) *Lognormal Distributions, Theory and Applications*. Marcel Dekker, New York.

Crump, K.S. (1998) On summarizing group exposures in risk assessment: is an arithmetic mean or a geometric mean more appropriate? *Risk Analysis*, 18(3): 293–297.

Dufour, A.P. (1984) Health effects criteria for fresh recreational waters, EPA-600/1-84-004. Office of Research and Development, US Environmental Protection Agency, Cincinnati, OH.

Esmen, N.A., and Hammad, Y.Y. (1977) Log-normality of environmental sampling data. *Journal of Environmental Science A*, 12(1–2): 29–41.

Ferley, J.P., Zmirou, D., Balducci, F., Baleux, B., Fera, P., Larbaigt, G., Jaco, E., Moissonnier, B., Blineau, A., and Boudot, J. (1989) Epidemiological significance of microbiological pollution criteria for river recreational waters. *International Journal of Epidemiology*, 18(1): 198–205.

Gilbert, R.O. (1987) *Statistical Methods for Environmental Pollution Monitoring*. John Wiley & Sons, Inc., New York.

Haas, C.N. (1983a) Effect of effluent disinfection on risks of viral disease transmission via recreational exposure. *Journal of the Water Pollution Control Federation* 55: 1111–1116.

Haas, C.N. (1983b) Estimation of risk due to low doses of microorganisms: a comparison of alternate methodologies. *American Journal of Epidemiology*, 118(4): 573–582.

Haas, C.N. (1996) How to average microbial densities to characterize risk. *Water Research*, 30(4): 1036–1038.

Institute for Environment and Health (2000) A review of the health effects of sea bathing water written by M.A. Mugglestone, E.D. Stutt, and L. Rushton, (Web Report W2). Institute for Environment and Health, Leicester, UK (http://www. le.ac.uk/ieh/webpub/webpub.html, posted November 2000).

Johnson, T., and Kotz, S. (1970) *Continuous Probability Distributions 1*. Houghton Mifflin, Boston.

Kay, D., Fleisher, J.M., Salmon, R.L., Jones, F., Wyer, M.D., Godfree, A.F., Zelenauch-Jacquotte, Z., and Shore, R. (1994) Predicting likelihood of gastroenteritis from sea bathing: results from randomized exposure. *Lancet*, 344: 905–909.

Kay, D., Bartram, J., Prüss, A., Ashbolt, N., Wyer, M.D., Fleisher, J.M., Fewtrell, L., Rogers, A., and Rees, G. (2004) Derivation of numerical values for the World Health Organization guidelines for recreational waters. *Water Research*, 38: 1296–1304.

Landwehr, J.M. (1978) Some properties of the geometric mean and its use in water quality standards. *Water Resources Research*, 14(3): 467–473.

Lehmann, E.L. (1983) *The Theory of Point Estimation*. John Wiley & Sons, Inc., New York.

Medema, G., van Asperen, I., Klokman-Houweling, J., Nooitgedagt, A., van de Laar, M.J.W., and Havelaar, A.H. (1995) The relationship between health effects in triathletes and microbiological quality of freshwater. *Water Science and Technology*, 31(5–6): 19–26.

National Technical Advisory Committee (1968) *Water Quality Criteria*. Federal Water Pollution Control Administration, Dept. of the Interior, Washington, DC.

Parkhurst, D.F. (1998) Arithmetic versus geometric means for environmental concentration data, *Environmental Science and Technology*, 32(3): 92A–98A.

Parkhurst, D.F., and Stern, D.A. (1998) Determining average concentrations of *Cryptosporidium* and other pathogens in water. *Environmental Science and Technology*, 32(21): 3424–3429.

Prieto, M.D., López, B., Juanes, J.A., Revilla, J.A., Llorca, J., and Delgado-Rodríguez, M. (2001) Recreation risks in coastal waters: health risks associated with bathing in sea water. *Journal of Epidemiology and Community Health*, 55(6): 442–447.

Seixas, N.S., Robins, T.G., and Moulton, L.H. (1988) The use of geometric and arithmetic mean exposures in occupational epidemiology. *American Journal of Industrial Medicine*, 14: 465–477.

Seyfried, P.L., Tobin, R.S., Brown, N.E., and Ness, P.F. (1985) A prospective study of swimming-related illness II. Morbidity and the microbiological quality of water. *American Journal of Public Health*, 75(9): 1071–1075.

US Environmental Protection Agency (1980) Final rule. *Federal Register*, 45: 33448 (May 19) (to be codified at 40CFR pt. 122).

US Environmental Protection Agency (1976) Quality criteria for water. US EPA, Washington, D.C.

US Environmental Protection Agency (1986) Ambient water quality criteria for bacteria – 1986, EPA 440/5-84-002. Washington, DC.

US Environmental Protection Agency (2003) EPA administered permit programs: The National Pollutant Discharge Elimination System, 40CFR122.41(l)(4)(iii), p. 210.

van Dijk, P.A.H., Lacey, R.F., and Pike, E.B. (1996) *Health Effects of Sea Bathing – Further Analysis of Data from UK Beach Surveys. Final Report to the Department of the Environment*. Report no. DoE 4126/3, Medmenham, UK.

Wade, T.J., Calderon, R.L., Sams, E., Beach, M., Brenner, K.P., Williams, A.H., and Dufour, A.P. (2006) Rapidly measured indicators of recreational water quality are predictive of swimming associated gastrointestinal illness. *Environmental Health Perspectives*, 114(1): 24–28.

Wymer, L.J., Brenner, K.P., Martinson, J.W., Stutts, W.R., Schaub, S.A., Dufour, A.P. (2005) The EMPACT Beaches Project: Results from a study on microbiological monitoring in recreational waters, EPA 600/R-04/023. Cincinnati, OH.

7

The EMPACT beaches: a case study in recreational water sampling

Larry J. Wymer

National Exposure Research Laboratory, US Environmental Protection Agency, Cincinnati, OH

7.1 Introduction

Various chapters describe sample and experimental design, use of the geometric or arithmetic mean, modeling and forecasting, and risk assessment in relation to monitoring recreational waters for fecal indicators. All of these aspects of monitoring are dependent on the spatial and temporal distribution of fecal indicator bacteria in the water. Knowledge of the distribution of indicator bacteria in water is particularly important in sampling design for monitoring. The US Environmental Protection Agency (EPA) conducted intensive water sampling studies during the summer of 2000 at five beaches in order to characterize temporal and spatial distribution of fecal indicator bacteria within the bathing areas of these beaches (Wymer *et al.*, 2005, hereafter referred to as the EMPACT report). Study beaches encompassed a variety of environments (Table 7.1).

Statistical Framework for Recreational Water Quality Criteria and Monitoring   Edited by Larry J. Wymer
© 2007 John Wiley & Sons, Ltd

Table 7.1   The EMPACT study beaches.

Beach	Location	Water body	Environment	Indicator bacteria
Belle Isle	Detroit, MI	Detroit River	Riverine	*E. coli*
West Beach	Ogden Dunes, IN	Lake Michigan	Freshwater	*E. coli*
Wollaston Beach	Quincy, MA	Quincy Bay	Marine	Enterococci
Imperial Beach	Imperial Beach, CA	Pacific Ocean	Marine	Enterococci
Miami Beach Park	Bowley's Quarters, MD	Chesapeake Bay	Estuarine	Enterococci

During the months of July and August 2000, water samples were collected at least twice daily from each of nine 'fixed' locations in the water, as determined by a *transect* and *zone* (Figure 7.1). A transect consisted of an imaginary line through a fixed point on the beach perpendicular to the shoreline. A zone (or 'depth zone') was defined as a contour line of equal water depth. As illustrated in Figure 7.1, each sampling location was defined by the intersection of transect and zone on a grid comprising three transects and three zones projected on the water's surface. A random point along the shoreline within the recognized beach area was selected to define the first transect (the leftmost transect in Figure 7.1). The middle transect was then determined as the parallel to the first transect at a distance of 20 meters, and the remaining transect as the parallel to the middle transect at an additional distance of 20 meters, or 40 meters from the first transect.

Locations at which the water attained a constant depth of 0.15, 0.5, and 1.3 meters (1.0 meters at Belle Isle, since buoys demarcating the swimming area were located at approximately this depth), corresponding to ankle-deep, knee-deep, and chest-deep water, were selected as sampling zones. Because three of the beaches were affected by ocean tides, the actual geographic locations of these zones varied according to the tide stage. Hence, sampling locations at these beaches were fixed only in the sense that they corresponded to locations at which the water depth was constant. Note the use of the term 'zone' rather than 'depth' in referring to areas of different water depth. This is to avoid confusion between water depth, which defines the sampling locations in Figure 7.1, and sampling depth, the depth below the surface of the water from which the samples were taken. In knee-deep and chest-deep water, the sampling depth was nearly always 0.3 meters from the surface, which is approximately 'face deep' for a swimmer. A relatively small number of samples were taken from other sampling depths as part of a parallel study. Zones may be considered as roughly corresponding to different activities among bathers, mainly wading and infant or toddler bathing (ankle-deep water), play (knee deep), and swimming or diving (chest deep).

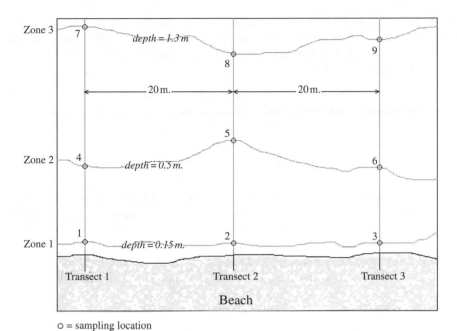

o = sampling location

Figure 7.1 EMPACT study beach sampling scheme.

7.1.1 Sampling schedule

On each day during July and August 2000, one of four sampling plans, referred to as 'basic', 'hourly', 'replicate', and 'depth' sampling, was followed. The total number of sampling visits and total samples collected under each of these sampling plans are summarized in Table 7.2 for each beach. The following are descriptions of each sampling plan.

Table 7.2 Number of sampling visits and total number of samples collected by type of sampling visit (from Wymer *et al.*, 2005).

Location	Basic sampling		Hourly		Replicate		Depth	
	Visits	Samples	Visits	Samples	Visits	Samples	Visits	Samples
West Beach	69	610	139	1162	16	416	8	144
Belle Isle	75	671	80	705	16	334	8	138
Wollaston	71	638	138	1242	16	407	8	144
Imperial Beach	68	612	140	1254	16	416	8	144
Miami Beach Park	72	643	140	1257	16	416	8	144

- *Basic.* On each of 53 basic sampling days, sampling was performed on two separate occasions, at 9:00 a.m. and 2:00 p.m. A single sample was collected at each of the nine sampling locations (1 through 9 in Figure 7.1) on each sampling visit.

- *Hourly.* On each of 21 hourly sampling days, sampling was performed on ten separate occasions, hourly on the hour, the first set being collected at 9:00 a.m. and the last at 6:00 p.m. A single sample was collected at each of the nine sampling locations on each sampling visit.

- *Replicate.* On each of 12 replicate sampling days, sampling was performed on two separate occasions, at 9:00 a.m. and 2:00 p.m. On each sampling visit, ten replicate water samples were collected at the central sampling location (location 5 in Figure 7.1), and two replicate samples were collected at each of the eight remaining locations. Because of a belief that river currents would make true replicate sampling difficult, the river site (Belle Isle) was an exception to the general replicate sampling scheme, with only two samples being collected at the central sampling point during the first four visits using the replicate sampling plan. Subsequent replicate sampling visits at Belle Isle reverted to the ten-replicate scheme after it was determined that currents were no more a factor at this sheltered beach than they were at the other beaches.

- *Depth.* On each of six depth sampling days, sampling was performed on two separate occasions, at 9:00 a.m. and 2:00 p.m., with samples being collected at each of the nine sampling locations in Figure 7.1 as on a basic sampling visit. In addition to the standard sampling depths (0.3 meters for knee- and chest-deep water) however, at each of the three sampling points in the knee-deep zone additional samples were collected at a depth of 0.425 meters (0.075 meters, or 3 inches, from the floor). At each of the three chest-deep zone sampling points, two additional samples were collected, at depths of 0.65 meters (half depth) and 1.225 meters (0.075 meters, or 3 inches, from the floor). Thus, a total of nine additional samples were collected beyond the nine samples collected in the basic sampling plan.

At each beach the actual number of days under which a given sampling plan was used may deviate from that given above because of inevitable circumstances, such as foul weather or rough water, that lead to deviations from the plan.

7.1.2 Microbiological analysis

Freshwater samples (Belle Isle and West Beach) were assayed for *Escherichia coli*, while the marine (Wollaston and Imperial Beaches) and estuarine (Miami Beach Park) water samples were analyzed for enterococci. In either case, membrane filtration methods are used (US EPA, 1997, 2000) which entail filtering volumes of samples through membranes, placing the membranes on a selective growth medium (agar), and incubating these for 22–24 hours under at controlled temperature. Resulting colonies of the targeted bacteria are then counted and assumed to have grown from a single 'colony-forming-unit' (CFU). US EPA (1986) criteria for fresh recreational waters specify either enterococci or *E. coli* as fecal indicator bacteria, *E. coli* being the most commonly used in practice. Only enterococci are recommended in the US EPA criteria for marine waters, since *E. coli* does not survive long in a saltwater environment. *E. coli* are generally present in greater numbers in freshwater than are enterococci in either freshwater or marine water. This is reflected in the US EPA (1986) criteria geometric mean limits for five samples or more over a 30-day period of 126 per 100 ml for *E. coli* in freshwater and 33 or 35 per 100 ml for enterococci in freshwater or marine water, respectively.

7.2   Probability mass function of indicators in water

7.2.1   Test for random dispersion

Ideally, indicators would be randomly dispersed throughout the water. By 'randomly dispersed' is meant that, (1) subject to an overall mean density, the number of organisms that would be found in some given volume of water cannot be predicted from (i.e., is independent of) the number found in some other non-overlapping volume, (2) the probability of finding a single organism is proportional to the volume (in the limit for some small volume), and (3) the probability of finding two or more organisms can be made to be negligible compared to the probability of finding a single organism by taking a sufficiently small volume of water. These are the three postulates from which the Poisson distribution is derived (Johnson *et al.*, 2005).

A random (Poisson) distribution of organisms would be the ideal because, due to the implied independence among different volumes of water, more intensive random sampling could be accomplished simply by taking larger volumes of water from the same location in the water. This means that the water is 'well mixed' or homogeneous. Two volumes of water collected in close proximity will not be any more correlated than two volumes collected a large distance away. Because of this, taking one large sample from one location in the water is equivalent to taking several samples from randomly selected locations. Beach managers and regulators would love an affirmative answer to the question 'Can I just stand in one spot and keep dipping my bucket to get more samples?'. Data from the EMPACT study can be used to provide an answer.

The Poisson probability mass function (pmf), $P(n) = e^{-\mu}\mu^n/n!$, gives the probability that that n events will occur. In our case this is the probability that n indicator organisms will be found in some volume, when the mean number of organisms per unit volume is μ. The variance of the Poisson pmf is equal to its mean (μ), a fact which provides a test of goodness-of-fit of sample results to the Poisson pmf. Fisher's index of dispersion (D^2) is the ratio of the sample variance to the sample mean multiplied by the degrees of freedom for the variance ($N - 1$), being a chi-square variable with $N - 1$ degrees of freedom. Fisher's D^2 can also be viewed as the equivalent of a chi-square goodness-of-fit test, with expected value equal to μ for each observation.

In order to show that collecting all samples from a single location in the water will work, we will have to show that the water is well mixed (i.e., Poisson distributed) across all three transects within a zone (we already know from the EMPACT study, and will show later in this chapter, that there are systematic differences across zones themselves). We may test for the randomness of results individually at each sampling visit using D^2 with two degrees of freedom. The individual D^2 may be added over all visits to the respective beach, given the additive property of chi-square variates, for an overall test for the joint random dispersion at that beach.

For each sample collected as part of the EMPACT study, aliquots of 100, 10 and 1 ml were assayed in order to accommodate those samples that will yield too many colonies to be counted from 100 ml as well as those samples that will yield only a small number of colonies in 100 ml. Thus, the typical sample volume used is 111 ml. However, in those cases in which 100 ml, and even 10 ml, aliquots were not countable, the usable volume will be less than 111 ml. Therefore, we consider the *rate* form of the Poisson distribution, $P(n) = e^{-\lambda v}(\lambda v)^n/n!$, in which λ is the concentration, or density, per unit volume (by convention we consider this to be per 100 ml). Fisher's D^2 is modified accordingly, and in fact is still simply a chi-square goodness-of-fit test with expected value for each cell equal to $v \cdot l$, where l is the value of λ as estimated from the aggregate counts and volumes from all replicate samples.

Results from these tests for random dispersion are presented in Table 7.3. That the indicator counts will 'pass' the test for random dispersion is obviously quite uncertain. Although the test appears to be non-significant more frequently for lower average indicator densities, this may be due to the fact that the test is weaker for lower average counts. Although not shown in Table 7.3, any compound test for a random distribution across all visits by depth zone and/or site will result in p-values less than 0.001, indicating highly significant departure from the Poisson distribution. The verdict is clear – another model describing the dispersion of indicator densities in water is required. Unfortunately, our beach sampler cannot stand at one location in the water and expect to obtain a 'representative' sample for the beach as a whole.

7.2.2   Overdispersed data: the negative binomial distribution

Failure of the Poisson distribution to apply to these data implies that the occurrences of indicator organisms in water are not independent or homogeneous. This could be

Table 7.3   Tests for random dispersion (Poisson distribution) within depth zone from the EMPACT study.

Beach (indicator organism)	Water depth	Average density (CFU/100 ml)	Total visits	Rejected as Poisson[a]	
				Number of visits	% of all visits
West Beach	Ankle	41	234	121	52
(*E. coli*)	Knee	29	233	72	31
	Chest	21	199	47	24
Belle Isle	Ankle	75	179	146	82
(*E. coli*)	Knee	70	179	149	83
	Chest	12	179	56	31
Wollaston Beach	Ankle	63	233	167	72
(enterococci)	Knee	41	233	140	60
	Chest	22	233	123	53
Imperial Beach	Ankle	11	231	85	37
(enterococci)	Knee	8	231	64	28
	Chest	5	231	55	24
Miami Beach	Ankle	105	236	205	87
Pk. (enterococci)	Knee	88	236	197	83
	Chest	19	236	109	46

[a] *P*-value for Fisher's D^2 (chi-square with 2 degrees of freedom) < 0.05.

interpreted as a mixture of Poisson distributions, that is, with varying values for λ. The distribution that is most often used to describe this situation is the negative binomial, which has the form

$$P(y) = \left(\frac{\kappa\mu}{1+\kappa\mu} \right)^{y} (1+\kappa\mu)^{-1/\kappa} \frac{\Gamma(y+1/\kappa)}{y!\Gamma(1/\kappa)}, \qquad (7.1)$$

and gives the probability of y occurrences in a fixed volume. $\Gamma(\cdot)$ represents the gamma function, which can be viewed as an extension of the factorial operator to non-integer numbers; for integer values of n, $n! = \Gamma(n+1)$. The mean of this distribution is μ and the variance is given by $\sigma^2 = \mu + \kappa\mu^2$. When $\kappa = 0$, the negative binomial pmf is the same as the Poisson. A negative binomial pmf will apply when the variation of the parameter of the Poisson pmf, λ, is described by the gamma distribution. The gamma distribution, a continuous probability density function, affords some flexibility in representing the distribution of the Poisson parameter, in that it can assume a range of shapes from nearly symmetric to highly right-skewed.

El-Shaarawi *et al.* (1981) found the negative binomial to adequately describe the distribution of bacterial densities in samples collected at different locations and over time in surface water.

The ten-replicate sampling events that were conducted in the EMPACT study afford us an opportunity for evaluating how well the Poisson and negative binomial pmfs as describe the distribution of water quality indicator bacteria. Table 7.4 shows that, similar to the duplicate samples, among the 70 sampling events when ten samples were collected from the same location, Fisher's D^2 indicated a decidedly non-Poisson distribution of values on 37 of those occasions. Overdispersion of the form $\sigma^2 = \mu + \kappa\mu^2$ may be specifically tested via a procedure from Dean (1992) using the statistic

$$P'_B = \frac{\sum \left[(Y_i - \hat{\mu}_i)^2 - Y_i - \ln(\hat{\mu}_i) \right]}{\left[2 \sum \hat{\mu}_i^2 \right]^{1/2}} \qquad (7.2)$$

where Y_i is the ith observation and $\hat{\mu}_i$ is its expected value (mean of all observations if the volumes are identical). This confirms the overdispersion on all 37 occasions. Deviance for the negative binomial model indicates how far the model is from the 'most likely' model. (By 'most likely' is meant the model that maximizes the likelihood function, which, in our case, would have each single observation as a Poisson variable associated with its own unique mean.) This deviance divided by residual degrees of freedom (9 in all cases for these data) is often used as a measure of how close the model is to the ideal, with a value of 1 indicating a perfect fit. Standard deviances resulting from the negative binomial model on the ten-replicate sampling results from each beach are shown in Table 7.5 and contrasted with the standard deviance obtained from the Poisson model. For these data, the average

Table 7.4   Tests for random dispersion (Poisson distribution) within ten-replicate samples from the EMPACT study.

	Ten-replicate sampling visits		Overall chi-square	Degrees of freedom	Overall p-value[b]
	Total no. of Visits	No. for which $p_0 < 0.05$[a]			
West Beach	16	9	597.26	144	< 0.0001
Belle Isle	6	6	559.61	54	< 0.0001
Wollsaton Beach	16	10	997.42	144	< 0.0001
Imperial Beach	16	6	431.39	144	< 0.0001
Miami Beach Park	16	6	688.75	144	< 0.0001

[a] Rejected as being Poisson distributed at $\alpha = 0.05$.
[b] Probability for the null hypothesis that *each* set of ten replicate samples is Poisson distributed.

negative binomial standard deviance is 1.17, very close to 1, and the maximum standard deviance is 1.74. The raw deviance is approximately a chi-square variate; thus the maximum deviance corresponds to a chi-square of 15.69 with 9 degrees of freedom, which is associated with a p-value of about 0.07, not at all significant especially considering that this is out of 70 such results.

The conclusion is that modeling fecal indicator bacteria in water samples with a Poisson distribution is not adequate. Thus, simply taking a larger volume of sample at one location in the water will not substitute for taking samples from other locations in the water. At least locally, when replicate samples are collected, the distribution of sample results has been shown to be suitably described by a negative binomial pmf.

Often, indicator bacteria counts are modeled as a lognormal distribution. The EPA's 1986 criteria for recreational water, for example, establish an upper limit which no single sample should exceed based on the 75th percentile of lognormally distributed indicator densities (EPA, 1986). Similarly, European Council Directives for recreational water quality (2006, Annex 2) are based in part on percentiles of a lognormal sampling distribution. One may ask how well the lognormal distribution describes these data if indicator bacteria CFU counts are truly distributed according a negative binomial pmf. Figure 7.2 compares the lognormal and negative binomial sampling distributions when the true mean, μ, is 50 CFU per 100 ml and the negative binomial dispersion parameter, κ, is 0.15, which is the average estimated parameter value among the replicate samples from the five EMPACT beaches (see Table 7.5). The variance based on $\kappa = 0.15$ is $\mu + \kappa\mu^2 = 394$. The true lognormal distribution curve is a continuous probability density function (pdf) rather than a discrete pmf like the negative binomial. However, the lognormal curve in Figure 7.2 is made to account for the discrete integer values of CFU counts by taking the

Table 7.5 Deviances for Poisson and negative binomial distribution models among ten replicates samples and negative binomial dispersion parameter estimates.

	Standard deviance[a]				k[b]
	Poisson average	Negative binomial			
		Average	Min.	Max.	
Belle Isle	9.962	1.361	0.934	1.419	0.072
West Beach	3.839	1.143	1.069	1.338	0.182
Wollaston Beach	6.109	1.183	1.081	1.454	0.257
Imperial Beach	2.687	1.235	1.022	1.743	0.227
Miami Beach Park	4.379	1.129	0.985	1.435	0.032

[a] The deviance from a generalized linear model for Poisson or negative binomial distribution (SAS PROC GENMOD) divided by its degrees of freedom.
[b] Negative binomial dispersion parameter ($\sigma^2 = \mu + \kappa\mu^2$).

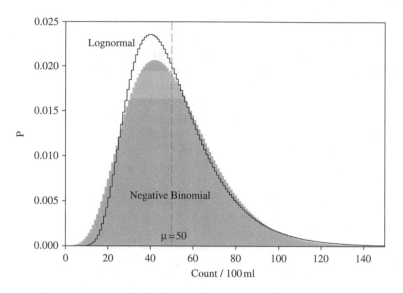

Figure 7.2   Comparison of lognormal and negative binomial probability distributions, both with mean 50 and variance 394. The lognormal curve has been 'discretized', giving the lognormal probability of an observation falling between $N - 0.5$ and $N + 0.5$.

probability of obtaining any given CFU count, N, as the area under the lognormal pdf between $N - 0.5$ and $N + 0.5$. It is evident that in this case the resulting lognormal distribution closely approximates the negative binomial.

7.3   Spatial and temporal influences of indicator bacteria densities

An important objective of the EMPACT bathing beaches study was to determine how fecal indicator bacteria densities varied in the water. Such information may have important implications for where and when to collect water samples. Spatial dimensions were defined in this study in terms of transects and depth zones. Temporal variation is in the form of day-to-day results, but also occurs between morning (9:00 a.m.) and afternoon (2:00 p.m.) samples. For our purposes, we will not consider samples that were collected from depths in the water other than 0.3 meters below the surface, since samples collected at other depths (0.5 meters below the surface and 0.15 meters from the bottom, depending on water depth) comprise such a very small fraction of the data. Also, we will not consider the hourly samples that were collected during certain weeks, other than those taken at 9:00 a.m. and 2:00 p.m. on those days.

Results from an analysis of variance (ANOVA) on the base 10 logarithms of the densities and assuming a lognormal distribution are detailed in the EMPACT report (Wymer *et al.*, 2005). Here, we will use the count data for CFUs directly, and assume that these follow a negative binomial pmf, as per the previous analysis. Generalized linear models (GLMs; McCullagh and Nelder, 1989) extend the traditional linear model, for example, the ANOVA model, to allow other probability distributions in addition to the normal distribution. Traditional linear models themselves belong to the class of GLMs, but the range is widened to encompass other analyses, including those for discrete data such as logistic and Poisson regressions.

Just as linear models require that observations follow a normal distribution, the GLM requires that observations follow a distribution that is a member of the exponential family. For the exponential family of distributions, the variance is a function only of (at most) the mean. Variance for the normal distribution is independent of the mean, being unity ($\sigma^2 = 1$) for the standard normal distribution, while the variance for the Poisson distribution is equal to its mean ($\sigma^2 = \mu$), as discussed earlier. For the negative binomial distribution, however, we saw earlier that the variance is expressed by $\sigma^2 = \mu + \kappa\mu^2$, and is thus dependent not only on the mean, μ, but also on an additional parameter, κ. Most statistical software for the estimation of GLMs will allow the negative binomial distribution by performing a separate estimate fort the negative binomial dispersion parameter, κ. For the following analysis, we use SAS/STAT version 9 software (SAS Institute, Cary, NC) PROC GENMOD.

Expected CFU counts from a plate will be proportional to the volume of sample that was filtered onto that plate. Although samples were collected in 500 ml polypropylene bottles, volumes of 100, 10 and 1 ml were actually filtered from each sample. On visits following the replicate sampling plan (see above), two or ten samples were collected from each location. These are accumulated in the GLM analysis so that the replicates from a single location would be counted as all of the CFUs enumerated from all of the filters produced from those replicates (i.e., a volume of as much as 1110 ml in total in the case of ten replicate samples). Because of the multiplicative effect, a logarithmic link function is used in the generalized linear model. The form of the model is given by

$$\ln(\mu_i) = \ln(V_i) + B_0 + \text{PM} + \text{Zone} + \text{Xsect} + \text{PM} \times \text{Zone} + \text{PM} \times \text{Xsect} \\ + \text{Zone} \times \text{Xsect} + \text{PM} \times \text{Zone} \times \text{Xsect} + \text{Date}. \tag{7.3}$$

Here, μ_i is the expected value of CFU count for the ith observation and V_i is the volume for that observation. In keeping with standard practice in measuring water quality indicator bacteria, volumes are expressed in terms of 100 ml equivalents. B_0 represents a constant (intercept) term, and PM, Zone, Xsect, and Date represent values that depend on whether the sample was from the morning or afternoon, what date the sample was taken, and the particular zone and transect from which the sample came. The 'crossed' terms, such as PM×Zone, represent interaction effects. The presence of a PM×Zone effect, for example, indicates differences in the effects of different depth zones depending on whether it is morning or afternoon. Observations of CFU counts, then, are assumed to come from a negative binomial distribution whose mean is given by equation (7.3).

A potential problem in applying model (7.3) to the EMPACT data is that observations may not be independent. At each sampling visit, samples were collected from nine locations in the water, each location being, in a sense, the same for each visit. At the two freshwater beaches the sampling locations were indeed from the same geographical locations at each visit, at least within the sample collectors' abilities to determine the sampling locations (Transects were determined by lines defined by fixed locations on the shore and samplers waded out to the proper depth as determined by measuring rods. There may have been inaccuracy in their determination due to wave action and sighting errors.) Among the East Coast and West Coast beaches and the estuarine beach, however, samples were not collected from the same geographical locations, but rather from the 'same' locations in the sense that they lie along the same transect and in water of the same depth each time. In either case, the same nine locations were used throughout the study, and samples from the same location may not be entirely independent. One could sum the results for each of the nine sampling locations, but this would result in a loss of information.

A method of handling correlated data like this is given by generalized estimating equations (GEE, Liang and Zeger, 1986). This technique explicitly accounts for correlation that may exist among data from a single unit that is measured repeatedly, such as is the case with sampling locations in the EMPACT study. Because these repeated measurements are made over time, it is reasonable to expect that the correlation structure will follow an autoregressive pattern. That is, if the correlation between successive measurements (consecutive points in time) is given by ρ, correlation between two measurements separated by n time periods is ρ^n. GEE estimation can be specified by identifying the sampling unit (location) and, in order to use this type of correlation structure, the within-location sampling unit (time). The basic model given by equation (7.3) does not change, only the computational method, which now accounts for the possibility that observations are not independent.

It may be adequate to disregard the fact that data consist of repeated observations at the same location if the correlation among the observations is negligible. In our case, given the fluid medium in which these measurements are made, one may suspect that this may be true. While it is never incorrect to take the correlation structure into account, even when the correlations are weak, there is some advantage to being able to disregard them if possible. Certain statistics resulting from a GLM maximum likelihood calculation, such as the likelihood ratio, are useful in evaluating alternative models. The GEE procedure does not use maximum likelihood and, thus, these statistics are not available.

Estimating the generalized linear model of equation (7.3) for the EMPACT data using maximum likelihood for independent data gives results that are, to varying degrees, similar to those obtained by using GEE. One expects to find the greatest difference between the two with respect to variance and associated standard errors of estimates, and among the five beaches these range from being nearly identical to twice as high when using GEE to take correlated measures into account. This suggests that approximate evaluations of main effect and interaction effect terms as a group can be made via likelihood ratio tests. These tests, similar in meaning to typical ANOVA tests, are shown in Table 7.6. Statistical significance or lack of such for the various effects is obvious, since the *p*-values tend to be unequivocal. Thus, there are few if any cases where one might suspect that ignoring the minor

Table 7.6   Likelihood ratio tests for main effects and interactions for the GLM.

Effect[a]	Degrees of Freedom	Likelihood Ratio	*p*-value	Effect	DF	Likelihood Ratio	*p*-value
Belle Isle				**Wollaston Beach**			
P[b]	1	201.38	< 0.001	P	1	16.93	< 0.001
Z[c]	2	1591.01	< 0.001	Z	2	310.64	< 0.001
T[d]	2	8.18	0.017	T	2	3.40	0.183
$P \times Z$	2	2.11	0.349	P×Z	2	11.03	0.004
$P \times T$	2	4.08	0.130	P×T	2	1.45	0.484
$Z \times T$	4	61.77	< 0.001	Z×T	4	1.06	0.900
$P \times Z \times T$	4	21.76	< 0.001	P×Z×T	4	2.04	0.727
Full model[e]	(924)	(569516.1)		Full model	(1010)	(350922.3)	
West Beach				**Imperial Beach**			
P	1	0.43	0.512	P	1	6.54	0.011
Z	2	295.20	< 0.001	Z	2	80.96	< 0.001
T	2	0.35	0.839	T	2	0.08	0.960
$P \times Z$	2	20.36	< 0.001	P×Z	2	3.54	0.170
$P \times T$	2	3.77	0.152	P×T	2	0.98	0.614
$Z \times T$	4	1.76	0.780	Z×T	4	2.60	0.626
$P \times Z \times T$	4	3.65	0.455	P×Z×T	4	3.32	0.506
Full model	(914)	(304366.9)		Full model	(992)	(61574.9)	
Miami Beach Park							
P	1	124.03	< 0.001				
Z	2	720.05	< 0.001				
T	2	14.05	< 0.001				
$P \times Z$	2	26.45	< 0.001				
$P \times T$	2	5.16	0.076				
$Z \times T$	4	16.55	0.002				
$P \times Z \times T$	4	1.93	0.748				
Full model	(993)	(1055795.5)					

[a] Marginal effect omitting the indicated factor or interaction from the full model.
[b] 2:00 p.m. vs. 9:00 a.m.
[c] Depth zone (ankle, knee or chest deep).
[d] Transect.
[e] Total degrees of freedom and −2 times the log-likelihood for the full model.

correlation effects may have led to a significant result. Differences among depth zones are seen to be highly significant in each of the EMPACT study beaches. Differences between the 9:00 a.m. and 2:00 p.m. results are also generally important. Only at Belle Isle and Miami Beach Park are means among transects seen to differ significantly.

Table 7.7   Generalized linear models for EMPACT data using generalized estimating equations

Parameter[a]	Complete model				Reduced model	
	Estimate[b]	Std. Error	Z	p_0^c	Estimate	p_0^c
Belle Isle						
Intercept	2.989				2.989	
PM[d]	−1.411	0.231	−6.108	<0.001	−1.411	<0.001
Zone 1	4.984	0.218	22.823	<0.001	4.984	<0.001
Zone 2	4.052	0.218	18.561	<0.001	4.052	<0.001
PM Zone 1	0.537	0.315	1.706	0.088	0.537	0.088
PM Zone 2	0.418	0.315	1.330	0.184	0.418	0.184
Xsect 1	−0.115	0.222	−0.519	0.604	−0.115	0.604
Xsect 2	−0.309	0.223	−1.390	0.164	−0.309	0.164
PM Xsect 1	0.952	0.321	2.966	0.003	0.952	0.003
PM Xsect 2	0.504	0.325	1.549	0.121	0.504	0.121
Zone 1, Xsect 1	0.369	0.309	1.195	0.232	0.369	0.232
Zone 1, Xsect 2	0.186	0.310	0.601	0.548	0.186	0.548
Zone 2, Xsect 1	−0.968	0.309	−3.133	0.002	−0.968	0.002
Zone 2, Xsect 2	−0.038	0.310	−0.124	0.901	−0.038	0.901
PM, Zone 1, Xsect 1	−1.544	0.442	−3.496	<0.001	−1.544	<0.001
PM, Zone 1, Xsect 2	−0.373	0.445	−0.840	0.401	−0.373	0.401
PM, Zone 2, Xsect 1	−0.630	0.442	−1.426	0.154	−0.630	0.154
PM, Zone 2, Xsect 2	−0.344	0.444	−0.775	0.439	−0.344	0.439
West Beach						
Intercept	2.164				2.910	
PM[d]	−0.341	0.201	−1.694	0.090	−0.357	0.003
Zone 1	0.681	0.195	3.500	<0.001	0.903	<0.001
Zone 2	0.101	0.196	0.515	0.606	0.200	0.079
PM Zone 1	0.848	0.279	3.042	0.002	0.596	<0.001
PM Zone 2	0.448	0.281	1.598	0.110	0.352	0.031
Xsect 1	−0.150	0.197	−0.759	0.448		
Xsect 2	−0.005	0.197	−0.023	0.982		
PM Xsect 1	0.039	0.284	0.137	0.891		
PM Xsect 2	−0.080	0.283	−0.282	0.778		
Zone 1, Xsect 1	0.495	0.275	1.801	0.072		
Zone 1, Xsect 2	0.166	0.275	0.604	0.546		
Zone 2, Xsect 1	0.244	0.278	0.880	0.379		

Zone 2, Xsect 2	0.061	0.277	0.221	0.825		
PM, Zone 1, Xsect 1	−0.580	0.394	−1.472	0.141		
PM, Zone 1, Xsect 2	−0.175	0.393	−0.445	0.656		
PM, Zone 2, Xsect 1	−0.305	0.398	−0.768	0.443		
PM, Zone 2, Xsect 2	0.011	0.397	0.029	0.977		

Wollaston Beach

Intercept	2.890				2.685	
PM[d]	−0.362	0.379	−0.957	0.339		
Zone 1	1.124	0.373	3.010	0.003	1.406	< 0.001
Zone 2	0.645	0.374	1.725	0.085	0.619	< 0.001
PM Zone 1	0.415	0.529	0.785	0.432		
PM Zone 2	−0.218	0.532	−0.411	0.681		
Xsect 1	−0.273	0.378	−0.723	0.470		
Xsect 2	−0.037	0.377	−0.099	0.921		
PM Xsect 1	0.320	0.536	0.597	0.550		
PM Xsect 2	−0.019	0.536	−0.036	0.971		
Zone 1, Xsect 1	0.154	0.529	0.291	0.771		
Zone 1, Xsect 2	0.107	0.528	0.204	0.839		
Zone 2, Xsect 1	0.163	0.530	0.307	0.759		
Zone 2, Xsect 2	0.047	0.529	0.090	0.929		
PM, Zone 1, Xsect 1	−0.428	0.749	−0.571	0.568		
PM, Zone 1, Xsect 2	−0.029	0.748	−0.038	0.969		
PM, Zone 2, Xsect 1	−0.281	0.753	−0.373	0.709		
PM, Zone 2, Xsect 2	0.220	0.752	0.293	0.769		

Imperial Beach

Intercept	1.657				1.701	
PM[d]	−0.115	0.236	−0.485	0.627	−0.215	0.007
Zone 1	0.797	0.228	3.491	< 0.001	0.767	< 0.001
Zone 2	0.467	0.230	2.030	0.042	0.249	0.010
PM Zone 1	−0.199	0.327	−0.609	0.542		
PM Zone 2	−0.286	0.330	−0.868	0.385		
Xsect 1	0.118	0.232	0.509	0.611		
Xsect 2	0.027	0.233	0.116	0.908		
PM Xsect 1	−0.177	0.333	−0.530	0.596		
PM Xsect 2	−0.203	0.334	−0.607	0.544		
Zone 1, Xsect 1	−0.329	0.323	−1.018	0.309		
Zone 1, Xsect 2	−0.049	0.323	−0.152	0.879		
Zone 2, Xsect 1	−0.366	0.326	−1.123	0.262		
Zone 2, Xsect 2	−0.108	0.325	−0.331	0.741		
PM, Zone 1, Xsect 1	0.762	0.462	1.651	0.099		
PM, Zone 1, Xsect 2	0.459	0.462	0.993	0.321		
PM, Zone 2, Xsect 1	0.280	0.468	0.598	0.550		
PM, Zone 2, Xsect 2	0.232	0.467	0.496	0.620		

Miami Beach Park

Intercept	3.878				3.794	
PM[d]	−1.310	0.269	−4.872	0.000	−1.502	< 0.001
Zone 1	2.535	0.261	9.707	0.000	2.838	< 0.001

Table 7.7   (Continued)

Zone 2	1.812	0.261	6.937	0.000	2.101	< 0.001
PM Zone 1	1.139	0.376	3.029	0.002	0.985	< 0.001
PM Zone 2	0.542	0.376	1.441	0.150	0.641	0.004
Xsect 1	−0.496	0.263	−1.884	0.060	−0.126	0.254
Xsect 2	−0.291	0.263	−1.109	0.267	−0.354	0.001
PM Xsect 1	−0.285	0.384	−0.742	0.458		
PM Xsect 2	−0.289	0.383	−0.755	0.451		
Zone 1, Xsect 1	0.877	0.370	2.371	0.018		
Zone 1, Xsect 2	−0.007	0.370	−0.018	0.986		
Zone 2, Xsect 1	0.802	0.370	2.166	0.030		
Zone 2, Xsect 2	0.080	0.370	0.217	0.828		
PM, Zone 1, Xsect 1	−0.416	0.535	−0.777	0.437		
PM, Zone 1, Xsect 2	0.062	0.534	0.115	0.908		
PM, Zone 2, Xsect 1	0.026	0.535	0.048	0.961		
PM, Zone 2, Xsect 2	0.307	0.534	0.575	0.566		

[a] Parameters are interpreted as coefficients of dummy variables. When the indicated condition or set of conditions is met, the estimated parameter value is added to the expected count.
[b] Natural logarithm of the indicator bacteria count per 100 ml.
[c] Probability under the null hypothesis that the respective parameter value is zero.
[d] 2:00 p.m.

Results from estimating the generalized linear model (7.3) using GEE are shown in Table 7.7. Day-to-day differences in levels of indicator bacteria are accounted for by the 'Date' parameter of model (7.3). Listing the separate date parameter estimates in Table 7.7 would add many rows to the table, given that visits were made on up to 62 days at each beach. Because our major concern here is with systematic differences that may exist among locations in the water and between morning and afternoon, the intercept term represents the average among all the days over which that beach was visited.

Parameter estimates are additive values (natural logarithms of indicator bacteria counts in 100 ml) that apply under the given conditions. This is the default dummy variable (0, 1) parameterization used in the SAS GENMOD procedure ('GLM' coding in SAS parlance) which makes it easy to determine expected values. In Belle Isle, for example, the expected natural logarithm of density in zone 1 (ankle-deep water) along transect 1 in the morning is $2.989 + 4.984 − 0.115 + 0.369 = 8.227$. Non-significant factors are eliminated for the reduced models in Table 7.7. If a main effect or interaction effect is significant in Table 7.6, all parameters associated with that effect are retained for the reduced model in Table 7.7. We are not so much interested in whether, for example, what we define as transect 1 differs from what we define as transect 2 as we are in making a statement that differences exist with regard to transects (locations along the shoreline) in general. The significance levels associated with the individual parameter estimates in Table 7.7 from the GEE analysis, however, are consistent with our earlier conclusions regarding overall main and interaction effects.

Whether the samples were taken in the morning or afternoon, and the zone and transect from which they were taken are seen to be significant determinants of indicator bacteria counts, although the specific main effect and interaction effects that were found to be important differ among the five different beaches. In general, the depth zone from which the sample was collected appears to be the single most important factor, with decreasing indicator densities the farther out from shore one goes. A smaller impact is seen from the fact that indicator densities were somewhat higher in general in the morning than they were in the afternoon. Bar charts of the modeled bacteria densities (from the full model given in equation (7.3) are shown in Figures 7.3, and graphically illustrate this effect.

The EMPACT report discusses the consistent differences in fecal indicator bacteria density as one moves farther from shore and the generally lower indicator levels in the afternoon, noting that the former may be influenced by stirring of the

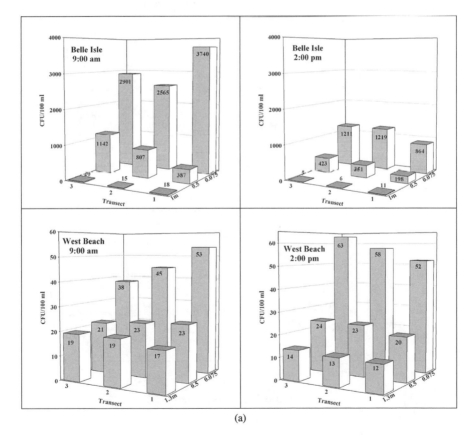

(a)

Figure 7.3   Predicted (a) *E. coli* and (b) enterococci CFU per 100 ml, 9:00 a.m. vs. 2:00 p.m., by transect and depth zone (water depth in meters).

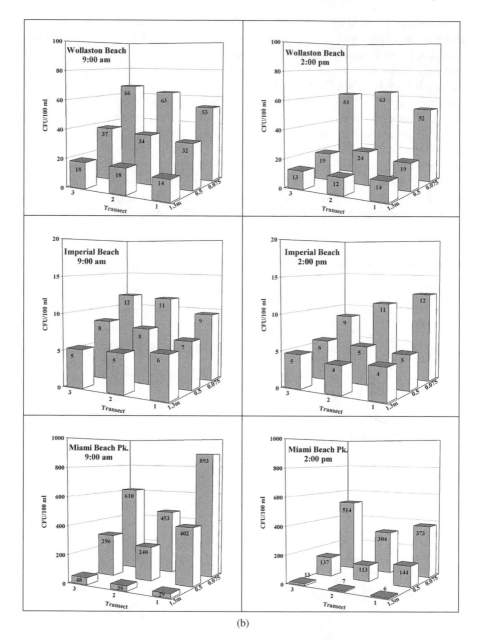

(b)

Figure 7.3   Continued

sediment and/or debris washing off the shoreline, while the latter is likely due to indicator bacteria die-off due to sunlight. Indeed, the morning to afternoon die-off was shown to be generally greater on sunny days. The differences among transects and the rather complicated interactions between transects, and depth zones, and morning-to-afternoon indicator levels in Belle Isle were posited by the EMPACT report to be a consequence of a small drainage ditch that emptied into the water nearly at the foot of the third transect. No similarly apparent influence was found at the Miami Beach Park that could explain the differences among transects seen at that site.

7.4   Components of variation

Developing a sampling plan for a bathing beach requires some estimate of the relevant sampling variances. The results of the previous section indicate that one may expect systematic differences to exist among different depth zones, with decreasing indicator density as one moves farther out from shore. This suggests a stratified sampling scheme, with depth zones as strata and subsampling within zones.

That differences between morning and afternoon indicator levels might also exist, as indicated by the previous section, suggests that these be treated as different populations. Bathers are likely to be present in the afternoon, not in the morning. However, methods of enumerating indicator bacteria which require growing colonies on plates of growth media, such as the membrane filtration methods in common use for this purpose, require a lengthy sample preparation and, especially, incubation time (about 24 hours) before yielding their results. The early morning is the most, and maybe the only, convenient time for collecting samples to be processed by membrane filtration. Even with state-of-the-art DNA detection, such as rapid (at present, 2-hour) polymerase chain reaction (PCR) techniques, or other rapid detection methods, morning sampling will be necessary in order to reach a timely decision in regard to posting or closing a beach.

Wade *et al.* (2006) reported on results from an epidemiological study, the National Epidemiological and Environmental Assessment of Recreational (NEEAR) Water Study, relating swimmers' illnesses to mean indicator levels for that morning as measured by a quantitative PCR method. Results established by the EMPACT study were used to design the sampling plan for the NEEAR study. Besides the importance of stratified sampling according to water depth and of selecting the time to sample, variance components as calculated from the EMPACT beaches enabled estimating the magnitude of sampling variance that might be encountered in the NEEAR study and, consequently, sampling errors corresponding to various sample size alternatives.

The data for these determinations are presented in Table 20 of the EMPACT report as variance components for $\log_{10}$ indicator counts per 100 ml among replicate samples ('pure error'), locations (transects) within zone, and among zones themselves. These as reproduced here in Table 7.8 with some additional supporting data in the form of the ANOVA results that lead to the estimates of the variance

Table 7.8   Estimated spatial variance components for $\log_{10}(\text{CFU}/100\,\text{mL})$ from the EMPACT study.

	Degrees of freedom	Mean square	E[mean square]	Estimated variance component
Belle Isle				
Visit	178	1.7507		
Zone(Visit)[a]	356	4.9693	$\sigma^2 + 1.0951\ \sigma_L^2 + 3.2569\ \sigma_z^2$	1.480
Location(Zone)[b]	1057	0.1510	$\sigma^2 + 1.1023\ \sigma_L^2$	0.103
Residual[c]	190	0.0374	σ^2	0.037
Total	1781			
West Beach				
Visit	233	2.0892		
Zone(Visit)	436	0.3418	$\sigma^2 + 1.0893\ \sigma_L^2 + 3.2531\ \sigma_z^2$	0.078
Location(Zone)	1332	0.0894	$\sigma^2 + 1.0995\ \sigma_L^2$	0.031
Residual	272	0.0554	σ^2	0.055
Total	2273			
Wollaston Beach				
Visit	232	3.0891		
Zone(Visit)	466	0.7915	$\sigma^2 + 1.0779\ \sigma_L^2 + 3.232\ \sigma_z^2$	0.203
Location(Zone)	1397	0.1353	$\sigma^2 + 1.0906\ \sigma_L^2$	0.036
Residual	263	0.0956	σ^2	0.096
Total	2358			
Imperial Beach				
Visit	231	1.3121		
Zone(Visit)	464	0.2288	$\sigma^2 + 1.0839\ \sigma_L^2 + 3.243\ \sigma_z^2$	0.017
Location(Zone)	1386	0.1724	$\sigma^2 + 1.0957\ \sigma_L^2$	0.032
Residual	272	0.1378	σ^2	0.138
Total	2353			
Miami Beach Park				
Visit	235	3.6317		
Zone(Visit)	472	3.1693	$\sigma^2 + 1.0827\ \sigma_L^2 + 3.235\ \sigma_z^2$	0.911
Location(Zone)	1408	0.2235	$\sigma^2 + 1.0945\ \sigma_L^2$	0.169
Residual	273	0.0391	σ^2	0.039
Total	2388			

[a] Variance component $= \sigma_z^2$.
[b] Variance component $= \sigma_L^2$.
[c] Variance component $= \sigma^2$ (pure error).

components. The ANOVA employs completely nested random effects, zones being considered as a random factor within each sampling visit (time), transects considered as random sampling 'locations' within zones, and replicate samples considered as random samples within a location (representing 'pure error'). Zones and transects were, of course, fixed throughout the study, and this biases variance estimates, particularly among zones where systematic differences (i.e., differences in general means) were shown to exist. Nevertheless, this gives an idea of the magnitude of variance along zones compared to the largely uncontrollable sources of variation, locations within zones and replicate samples, were one to ignore the systematic zone effects and fail to stratify the sampling plan accordingly.

The SAS GLM procedure, which uses least squares to fit linear models and assumes normality for tests of significance and confidence intervals, was used for this analysis. This is in contrast to GENMOD, for *generalized* linear models in the previous section which uses maximum likelihood estimation, sometimes in combination with the method of moments to fit linear models from a number of distribution types. The application of least squares presented here involves estimation only. Expected values for ANOVA mean squares considering locations and zones as random effects are shown in Table 7.8, and estimates resulting from these expectations are given in the last column.

These variance components are interpreted as the amount that is added to variance by including the corresponding dimension in the random sampling plan. In West Beach, for example, taking all samples from one spot in the water results in a variance (in $\log_{10}$ density per 100 ml) of 0.055. Taking these samples from different locations in the water instead (but from water of the same depth), increases variance to 0.086 (0.055 + 0.031). Because there are no systematic differences among transects at West Beach, the only way to account for the transect variation is to sample. Depth zones, however, did show significant systematic differences (differences in overall mean among depths). Thus, the variation among different zones can be accounted for by stratification. Note that at Imperial Beach the variance component among depth zones is small compared to pure error variance or the contribution of transects to sampling variance. Still, differences among depth zones at Imperial Beach are due to differences in general means among the depth zones, although the advantage of stratifying apparently is not as great as it would be at the other beaches.

7.5 Other findings from the EMPACT study

The EMPACT report discusses other details that will only be summarized here. Most notable, perhaps, were the large differences from day to day that were observed to occur at each of the beaches. As stated in that report, for each of the studied beaches between one day and the next the geometric mean indicator density (averaged over all nine sampling locations) changed by more than a factor of 2 (i.e., it doubled or halved in value) about half the time. With existing membrane filtration methods, there is little that can be done to take this source of variation into account, since these methods require a full day to yield their results. The philosophies behind the 1986 EPA criteria and the 2006 EU directive lean toward long-term process control, this being the only reasonable alternative using a 24-hour method. Rapid PCR methods, previously discussed, are a potential solution for providing more current information on water quality at a beach.

Modeling of effects due to weather, bathers, and tides was also presented in the EMPACT report. In short, rainfall, onshore winds, and incoming tides were consistently found to result in increasing indicator levels. Recreational water quality modeling is discussed in detail in the following chapters in this book.

References

Dean, C.B. (1992) Testing for overdispersion in Poisson and binomial regression models. *Journal of the American Statistical association*, 87: 451–457.

El-Shaarawi, A.H., Esterby, S.R., and Dutka, B.J. (1981) Bacterial density in water determined by Poisson or negative binomial distribution. *Applied and Environmental Microbiology*, 41(1): 107–116.

European Council (2006) Directive 2006/7/EC of the European Parliament and of the Council of 15 February 2006 concerning the management of bathing water quality and repealing Directive 76/160/EEC. *Official Journal of the European Union*, I64,4.3.06: 37–51.

Johnson, N.L., Kemp, A.W., and Kotz, S. (2005) *Univariate Discrete Distribution*. Wiley, Hoboken, NJ.

Liang, K.Y., and Zeger, S.L. (1986) Longitudinal data analysis using generalized linear models. Biometrika, 73: 13–22.

McCullagh, P., and Nelder, J.A. (1989) *Generalized Linear Models*, 2nd edn. Chapman & Hall, London.

US Environmental Protection Agency (1986) *Ambient water quality criteria for bacteria – 1986*, EPA 440/5-84-002. Washington, DC.

US Environmental Protection Agency (1997) Method 1600: Membrane filter test methods for enterococci in water, EPA-821/R-97/004a. Office of Water, Washington, DC.

US Environmental Protection Agency (2000) Improved enumeration methods for the recreational water quality indicators: enterococci and *Escherichia coli*, EPA/821/R-97/004, Office of Science and Technology, Washington, DC.

Wade, T.J., Calderon, R.L., Sams, E., Beach, M., Brenner, K.P., Williams, A.H., and Dufour, A.P. (2006) Rapidly measured indicators of recreational water quality are predictive of swimming associated gastrointestinal illness. Environmental Health Perspectives, 114(1): 24–28.

Wymer, L.J., Brenner, K.P., Martinson, J.W., Stutts, W.R., Schaub, S.A., and Dufour, A.P. (2005) The EMPACT Beaches Project: Results from a study on microbiological monitoring in recreational waters, EPA 600/R-04/023. Cincinnati, OH.

8

Microbial risk assessment modeling

Graham McBride

National Institute of Water and Atmospheric Research, Hamilton, New Zealand

8.1 Why do modeling?

Risk modeling is useful in setting health criteria or guidelines (Westrell *et al.*, 2004). However, some water managers, and scientists, can exhibit a rather dismissive attitude to its use in water-related microbial risk assessments, especially when it relies in some measure on statistical concepts. This attitude is characterized by a belief that the key patterns that modeling attempts to identify – associations between microbial contamination of recreational waters and the health of recreational water users, and (even better) their cause and effect paths – should be 'obvious'.[1]

There is, of course, a grain of truth in such assertions, particularly when effects are 'large' and the associated studies are 'small'.[2] But for microbial contamination

---

[1] Such statements are usually heard in conversations and in discussion of evidence presentations in legal settings, and so the very desirable convention of citing example sources of such statements is difficult. Claims that cause – effect paths are identified by mere associations between exposures and effects should always be assessed against the Bradford-Hill criteria (Hill, 1965; Doll, 1991).

[2] A large effect would have an attack rate of at least 10% and a small study (for a controlled cohort design) would have less than 500 participants (an uncontrolled study would still be 'small' with a larger number of participants). Such a study may not detect an effect (using statistical null hypothesis tests), even though that effect may appear to be 'obvious'.

Statistical Framework for Recreational Water Quality Criteria and Monitoring Edited by Larry J. Wymer
© 2007 John Wiley & Sons, Ltd

of recreational waters the effects are often 'small' and studies have to be 'large' to determine them.[3] In such studies it is extremely difficult – if not impossible – to determine the patterns 'by eye'. But when they are conducted, we routinely find that the health of recreational water users *is* affected by the microbial quality of the water they have contact with (via swimming, wind-surfing, . . .); see Prüss (1998). This finding can even be true for those who merely wade in recreational water, not immersing their head as a swimmer does.[4]

These findings also bear on a follow-on question commonly asked by skeptical colleagues: 'Where are all the sick swimmers?' The twofold answer is: (a) they are in the community, and (b) they can be only be identified by performing an exposure survey and some kind of modeling study, to assess the microbial health risk.

Two main approaches to microbial health risk assessment modeling may be used: epidemiological studies, and quantitative microbial risk assessment (QMRA). Both must be considered in relation to this chapter's topic. A brief history of their development and use is included, with the intention that it should highlight key issues that have been addressed and, in part, resolved. The chapter ends with a brief comparison of their merits.

8.2   Epidemiological studies

The essence of these studies is to:

- recruit a group of people who attend recreational waters, such as beaches, that have measured contamination levels;

- make enquiries concerning their subsequent health status; and

- examine possible relationships between the degree of contamination and these health effects.

The follow-up enquiry typically occurs 3–5 days after exposure, and sometimes also at a longer interval (e.g., 3 weeks; Kay *et al.*, 1994). Ideally, some of the beach attenders should be those who do not enter the water at all, to establish a baseline for comparisons.[5] Some studies are retrospective, examining historical records of

---

[3] By 'small' effects, we mean that the illnesses that may be contracted are generally mild (typically 4–8 days; Fleischer *et al.*, 1998), and their attack rates are low (of the order of, at most, a few percent of exposure occasions). They also are often not reported or recorded. But note that, from a public health point of view, this does *not* imply that the health effects are minor – in any year a substantial proportion of a country's population are exposed to recreational waters, mostly on multiple occasions, and so the number of illness cases can be significant, even though the attack rate may be low on any one occasion. Furthermore, serious sequelae may arise in a small proportion of cases (Helms *et al.*, 2003), such as Guillain–Barré syndrome following onset of campylobacteriosis (Nachamkin *et al.* 2000).

[4] Respiratory illness may be contracted from inhalation of aerosols by wading adults tending children playing in surf near the shore; the aerosols derive from wave action (McBride *et al.*, 1998).

[5] But note that respiratory infections caused by aerosolized pathogens can affect *all* beach attenders (Baden *et al.* 1992; Chang *et al.* 2001).

illness and exposure, but these days most are prospective in which participants are contacted at the beach on the day of exposure and later followed up for illness data.

Two further distinctions between study types are relevant:

- Whether the groups are 'controlled' or 'uncontrolled'. Uncontrolled groups are people who have chosen to attend a beach having minimal knowledge (if any) that a study is being conducted, while controlled groups are explicitly recruited for the study before attending the beach, and their exposures (including food intake) are controlled. This also permits the very desirable attribute of randomization of swimmers and non-swimmers.

- Whether individuals with multiple exposures between the exposure day and subsequent follow-up are included. Multiple exposures are inevitable in a 'longitudinal study', in which a group is followed over time (e.g., Calderon *et al.*, 1991), but they may also feature in a 'cross-sectional study' (e.g., Stevenson, 1953) in which the exposure of interest occurs at a single time (in Stevenson's study further exposures occurred outside that time).

Examples of such studies are given in Table 8.1.

The need and basis for water quality criteria for recreational waters in the 1960s and 1970s was characterized by strong debate. The results of a retrospective case – control study for British seaside towns reported by the Public Health Laboratory Service (PHLS, 1959) were interpreted by the Medical Research Council (1959) to mean that 'with the possible exception of a few aesthetically revolting beaches round the coast of England and Wales, the risk to health of bathing in sewage contaminated sea water can, for all practical purposes, be ignored'. This inference, and the somewhat inconclusive results from large prospective studies reported by Stevenson (1953), led some to argue that there was no case for microbiological recreational water quality criteria (e.g., Moore 1975), even though some had already

Table 8.1 Five main types of epidemiological studies used to develop recreational water quality criteria and standards.

Type	Important examples
Retrospective	
Case – control	PHLS (1959)
Follow-up	Ferley *et al.* (1989)
Prospective	
Uncontrolled, multiple exposures	Stevenson (1953), Calderon *et al.* (1991)
Uncontrolled, single exposure	Cabelli (1983), Dufour (1984), Wade *et al.* (2006), Colford *et al.* (2007)
Controlled cohort, single exposure	Kay *et al.* (1994), Fleisher *et al.* (1996)

been set (e.g., National Technical Advisory Committee 1968). Others noted that the British case – control study referred to above was for two very serious illnesses only – enteric fever (paratyphoid or typhoid) and poliomyelitis – whereas the bulk of illness was thought to be of a much wider range of milder illnesses. Also, the design of the Stevenson studies was faulted, for example, by not including head-immersion in the definition of swimmers, and for allowing multiple exposures (Cabelli *et al.*, 1975).

Resolution of many of such issues was achieved in the influential uncontrolled prospective studies of Cabelli (1983) for seawater, and by Dufour (1984) for freshwaters (lakes). These studies have been used as the basis for water quality criteria (e.g., US Environmental Protection Agency (EPA), 1986). They too have been criticized, particularly for the lack of control of the studies' subjects, their statistical modeling methods, and their lack of analysis of respiratory illness effects. Alternative controlled cohort studies and logistic modeling methods have been conducted in England and Wales, identifying stronger relationships between fecal contaminations of recreational waters and health effects of swimmers, both gastrointestinal (Kay *et al.*, 1994) and respiratory (Fleisher *et al.* 1996).

Taken as a whole, studies since the late 1970s (when Cabelli's studies were performed) demonstrate that swimmers' health *is* affected by the quality of the recreational water when that quality is indexed to appropriate bacterial indicators – intestinal enterococci for seawater and also *E. coli* for freshwater (Cabelli, 1989; Prüss, 1998; Wade *et al.*, 2003). The controlled cohort studies now form the basis of revised World Health Organization guidelines (WHO, 2003) and for some other countries as well (Ministry for the Environment and Ministry of Health, 2003; National Health and Medical Research Council, 2005). Their derivation is explained in detail by Wyer *et al.* (1999) and Kay *et al.* (2004).[6] In essence, the approach is to grade beaches on the basis of their historical microbiological water quality data *and* on sanitary surveys of the beach watershed's susceptibility to fecal contamination. This has the feature of enabling beach management to rely less on water quality data and to make better use of environmental pollution predictors, as described for beaches in Sydney, Australia (Ashbolt and Bruno, 2003).

8.2.1   Example study

Coastal recreation activities are commonplace in New Zealand – swimming, yachting, skiing, and various forms of surfing. This popularity, and the results of a preliminary study, provided the stimulus for a multi-agency study of health effects of marine bathing at seven beaches (full details are given by McBride *et al.*, 1998). All necessary ethical approvals were obtained.

The beaches were classified into three groups: pristine (2 beaches), predominant animal fecal impact (2 beaches), and predominant human fecal impact (3 beaches).

---

[6] Modifications have recently been suggested (Wymer *et al.* 2005).

Beachgoers were approached by trained interviewers, and eligible persons were followed up 3–4 days later to ascertain any subsequent illness (persons who also attended beaches 3 days prior to or after their beach interview were ineligible). Illness categories included gastrointestinal and people-related illness categories (ear, eye, throat infections, or skin rash), and also respiratory illness. After consultation with health professionals, the definition of this respiratory illness category included non-febrile respiratory symptoms (other studies confined attention to febrile respiratory symptoms, e.g., Fleisher *et al.*, 1996).[7] A total of 107 beach-day surveys were carried out over a summer season, during which subsurface water samples were collected twice daily at three or four locations along the beach from both knee-depth and chest-depth water and analyzed for enterococci, *E. coli*, and fecal coliforms. Samples for each depth and time were pooled before laboratory analysis, giving four results per beach-day (two depths and two times). Data for the two times were pooled prior to statistical analysis, reflecting individuals' swimming occasions on the exposure day. The degree of contamination at the beaches was relatively low compared with historical data and with the current microbiological water quality guidelines. Concentrations in the knee-depth water were consistently higher than in the deeper water.

Usable responses from 3884 individuals were obtained; 2307 were 'non-exposed' (did not enter the water), leaving 1577 in the 'exposed' group – 377 paddlers (did not immerse the head) and 1200 swimmers. Crude unadjusted risk differences between the exposed and non-exposed groups indicated potential health effects (Table 8.2), notably for long-duration swimmers (those with exposure in excess of $1/2$ hour) and also for paddlers (many of whom were adults looking after small children in shallow water).

Table 8.2 Crude risk differences (unadjusted) per 1000 individuals[a] in the New Zealand study.

Group	HCGI[b]	RESPI[c]
Paddlers	3.9	6.7
Swimmers	3.5	16.7
swam $< 1/2$ hour	0.3	1.5
swam $> 1/2$ hour	7.1	33.2

[a] Other illness categories omitted, in the interests of clarity (for full details see McBride *et al.*, 1998).
[b] Highly credible gastrointestinal illness, defined as any vomiting, or loose bowel with fever, or loose bowel with 'disability', or nausea with fever, or indigestion with fever ('disability' was defined as one or more days away because of illness, or days unable to do normal activities, or having sought medical advice, or having been hospitalized).
[c] Respiratory illness, defined as dry cough, or cough with spit, or shortness of breath, or chest pain with fever.

---

[7] Self-diagnosis of fever may fail; some significant respiratory illnesses are not accompanied by fever (e.g., responses to allergenic material resulting from microbial or fungal elements, asthma).

The degree to which these risks were influenced by faecal indicators was examined using generalized linear models (Armitage and Berry, 1987). Separate models were identified for each illness category. Previous studies (e.g., Cabelli, 1983) had used a form of ordinary linear regression, in which risk differences and logarithms of indicator concentrations were put in groups defined either by ranges of concentrations or by beach (e.g., Cabelli, 1983). As a result of this grouping, the final regression was performed on only a few data points (one per group). In contrast, no grouping is required when using generalized linear models because each datum enters the model individually, giving a superior performance. A number of forms of these models can be derived (McCullagh and Nelder, 1995); commonly, recreational water studies have used the logistic model (Seyfried *et al.*, 1985; Pike, 1994; Fleisher *et al.*, 1996; Wymer and Dufour, 2002). In this model the dependent variable on the left-hand side of the model's equation is the 'logit', defined as $\log_e[p/(1-p)]$, where p is the absolute illness risk. The right-hand side contains a linear mixture of the *exposure variable* (faecal indicator concentration, possibly including threshold values), *confounders* (e.g., age, gender, food intake), and *effect modifiers* (such as swim duration).[8] The illness distribution is assumed to be binomial – as one may expect for a binary variable (ill/not-ill). An attractive feature of this approach is that one can obtain the odds ratios for the model's risk factors, by simple exponentiation of the model equation's coefficients.[9]

However, for two main reasons, relative risks may be regarded as more appropriate measures than odds ratios. First, the former are better understood by the public at large (with the possible exception of some gamblers). Second, the odds ratio tends to create an inflated view of actual risk; at low values (less than 2) the difference is small, but at higher risks they begin to diverge substantially. For these reasons, this study did not use the logit transformation, using just the logarithm of p – a 'log-linear model'. Once again, the illness risk was assumed to follow a binomial distribution.

Attempting to fit a linear model directly can obscure the main patterns of association in the data, because threshold effects may occur. Therefore, each fecal indicator concentration was replaced by a categorical variable – the number of the quartile in which the concentration lay.[10] Therefore, the main results were expressed in terms of quartiles of indicator concentrations.

Of the three fecal indicators used, enterococci were the most strongly related to illness risks. Concentrations for either water depth, or their pooled results, were simi-

---

[8] Predicted risks from the model are 'adjusted' for confounders, but not for effect modifiers.

[9] Let the number of individuals in the exposed category who become ill be a, and those who fail to become ill be b. In the non-exposed category, denote the numbers of ill and not ill by c and d, respectively. The odds ratio of disease is $(a/b)/(c/d) = ad/bc$. The relative risk of disease is $[a/(a+b)]/[c/(c+d)] = a(c+d)/[c(a+b)]$. If the disease is rare (no more than a few percent) then $b >> a$ and $d >> c$, and so the odds ratio and relative risk ratio are very similar in value.

[10] Actual indicator concentrations can be used in a model once evidence is found for a progressive increase in risk through the quartiles.

Table 8.3 Relative risk of illness for groups exposed to enterococci quartiles, compared to non-exposed (age-adjusted).

Exposure class	Quartile 1	Quartile 2	Quartile 3	Quartile 4
HCGI[a]				
All exposed	1.13	0.84	1.00	1.38
Paddlers	0.52	0.67	2.13	2.30
Swimmers	1.28	0.88	0.76	1.06
swam $< 1/2$ hour	0.94	0.91	0.71	1.05
swam $> 1/2$ hour	1.69	0.86	0.81	1.08
RESPI[a]				
All exposed	0.77	1.18	1.40	2.91*
Paddlers	0.41	0.54	1.22	4.53*
Swimmers	0.86	1.30	1.43	2.46*
swam $< 1/2$ hour	0.71	1.08	0.53	1.59
swam $> 1/2$ hour	1.06	1.58	2.14	3.31*

[a] See footnotes to Table 8.2 for definition of these terms.
* Statistically significant (two-sided test) at the 5 % level (using the technique of Greenland and Robins, 1985).

larly associated with illness risks (consistent with swimmers and boogie-boarders in high-energy environments 'sampling' many depths during water contact).

Key results, in Table 8.3, show that at the relatively low levels of fecal contamination prevailing at the seven beaches during this study, highly credible gastrointestinal illness risk never attained statistical significance with enterococci concentrations, whereas respiratory illness did – in the top concentration quartile. These associations were particularly evident among paddlers and long-duration swimmers. Further results showed that respiratory illness risks were statistically significantly higher at the beaches impacted by human and animal fecal material, compared with the pristine beaches. Also, the risks at the two types of impacted beaches were not separable. Given the substantial agricultural component of the New Zealand economy, this latter finding has reinforced the notion that, at least in New Zealand, health risks from exposure to human versus animal fecal material should not necessarily be separated in water quality criteria and standards. This is an area deserving of further research, using both epidemiological and QMRA approaches.

8.3 Quantitative microbial health risk assessment

The focus in epidemiological studies is on *human subjects*, in their exposure to whatever pathogens may be present in the water. In contrast, the newer field of quantitative microbial risk assessment (QMRA) focuses on *particular pathogens* that humans may be exposed to. QMRA is described in considerable detail by Haas *et al.* (1999). It follows a four-step process:

1. *Hazard identification.* Which pathogens? (A hazard is a quantity that has a potential to cause harm.)

2. *Exposure assessment.* What exposure might a population have to pathogens via water contact recreation?

3. *Dose – response analysis.* What is the probability of infection (or illness) given ingestion or inhalation of one or more pathogen particles?

4. *Risk characterization.* How much infection or illness would arise in a population exposed to a distribution of pathogens in the water? What is the risk of illness faced by an exposed individual?

An example of a hazard identification exercise is given in Table 8.4.

The next two steps carry a common thread: some items characterizing exposure and dose response are inherently variable or uncertain, and so are described by statistical

Table 8.4   Potential pathogens in wastewater effluents and receiving waters: a New Zealand example (for Christchurch City coastal waters).

Pathogen	Main disease caused	Comments	Include?
Bacteria			
Campylobacter spp.	Gastroenteritis	Poor survival in seawater	No
Pathogenic *E. coli*	Gastroenteritis	Low concentration expected in sewage	No
Legionella pneumophila	Legionnaires' disease	No evidence of environmental infection route	No
Leptospira sp.	Leptospirosis	Low concentration expected in sewage	No
Salmonella sp.	Gastroenteritis	Low concentration expected in sewage	No
Salmonella typhi	Typhoid fever	Rare in New Zealand	No
Shigella sp.	Dysentery	Low concentration expected in sewage	No
Vibrio cholerae	Cholera	Rare in New Zealand	No
Yersinia enterolitica	Gastroenteritis	Low concentration expected in sewage	No
Helminths			
Ascaris lumbricoides	Roundworm	Rare in New Zealand	No
Enterobius vernicularis	Pinworm	Low concentration expected in sewage	No

Fasciola hepatica	Liver fluke	Rare in New Zealand	No
Hymnolepis nana	Dwarf tapeworm	Rare in New Zealand	No
Taenia sp.	Tapeworm	Rare in New Zealand	No
Trichuris trichiura	Whipworm	Rare in New Zealand	No
Protozoa			
Balantidium coli	Dysentery	Low concentration expected in sewage	No
Cryptosporidium oocysts	Gastroenteritis	Can accumulate in shellfish, but virus groups may be of more concern	No
Entamoeba histolytica	Amoebic dysentery	Rare in New Zealand	No
Giardia cysts	Gastroenteritis	Poor survival in seawater	No
Viruses			
adenoviruses	Respiratory disease[a]	Very infective, present in substantial concentrations in raw sewage	**Yes** (swimming only)
enteroviruses	Gastroenteritis	Less infective, but health consequences can be more severe than adenovirus	**Yes** (swimming and shellfish)
hepatitis A virus	Infectious hepatitis	Low sewage concentration; very infective Can affect surfers in contaminated waters[b]	**Yes** (shellfish)
noroviruses[c]	Gastroenteritis	No reliable method for viability enumeration; limited data on occurrence in water and infectivity.	No
rotaviruses	Gastroenteritis	Limited evidence of waterborne infection in NZ; infection in children would be of concern.[d]	**Yes** (swimming and shellfish)

Main source: ANZECC & ARMCANZ (2000); Carr (2001).
[a] Adenoviruses can also cause pneumonia, eye infections and gastroenteritis.
[b] Gammie and Wyn-Jones (1997).
[c] Formerly known as Norwalk-like viruses.
[d] A rationale for concern about potential contamination of shellfish by rotavirus has been given by Rose and Sobsey (1993).

distributions. Swimming is an example of the former – some people swim for longer periods than others and their ingestion and inhalation rates will differ. Each of these variables can be described by a distribution. Uncertainty is particularly evident in dose – response analysis, because it relies on limited clinical trial data reported in the literature, generally with only one strain of a pathogen species, whereas it may be that dose response can vary from one strain to another. Available dose – response models are summarized by Teunis *et al.* (1996) and Haas *et al.* (1999).

Applying averages to all the variable and uncertain quantities can give rise to misleading results. To avoid such problems Monte Carlo or bootstrap techniques are employed – although simpler arithmetic mean models may be used for linear dose – response models (Gale, 2003). These calculations are made for groups of people at an exposure site (a beach) on many occasions. For each of those occasions a random sample is taken from statistical distributions of the quantities that are either uncertain or variable. In this way the range of variability and uncertainties is explicitly accounted for.

The results of these calculations to be used in the last step (risk characterization) are twofold:

- a 'risk profile', giving the percentage of time that infection risks for the group of people are up to a stated value;

- the individual infection risk (IIR), being the risk that an individual faces if exposed to the beach water on a random day.

These calculations are made for infection, rather than illness, for two reasons. First, if the infection rate is minimized, then so also is the illness rate, whereas controlling only illness may fail to adequately contain infection (infected individuals may shed the pathogen into the environment, causing illness in others). Second, dose – response relationships established from clinical trials are much better defined for infection than is the case for illness, because some individuals may be infected yet asymptomatic, even at very high doses.

To date, the majority of the published applications of QMRA to aquatic systems have been for drinking-water (e.g., Haas *et al.*, 1996; Masago *et al.*, 2002; Makri *et al.*, 2004; Pouillot *et al.*, 2004), although Mena *et al.* (2003) address coxsackievirus risks for both drinking-water and recreational water. Crabtree *et al.* (1997) demonstrate an application of adenovirus infection and illness risks for both drinking-water and recreational water. A further example application of adenovirus infection for recreational waters (for a proposed ocean outfall of treated wastewater for Christchurch, New Zealand) is depicted in Figure 8.1.[11] Example results appear in Table 8.5. These results show the percentage of time that the incremental risk is up to the stated number – the number of adenovirus infections per 1000

---

[11] Adenovirus has become of particular concern because it is very infectious (median infective dose is less than two particles – Haas *et al.*, 1999), it is present in significant concentrations in raw sewage (Jacangelo *et al.*, 2003), and there is concern that at least some strains may be very resistant to disinfection (Thompson *et al.*, 2003).

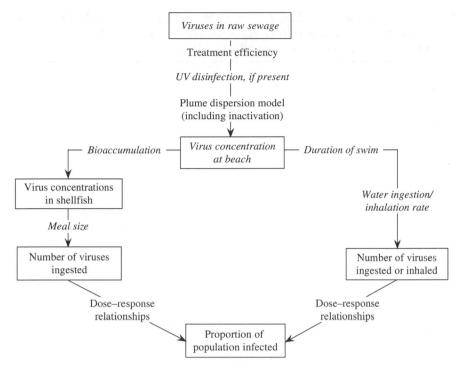

Figure 8.1   General flow diagram for QMRA calculation sequence for swimmers and for consumers of raw shellfish in Christchurch coastal waters. Italicized text dcnotcs variablcs for which random samplcs arc takcn from thcir assigncd statistical distributions.

person-swimming events. The IIR also shown in the table represents the risk to a swimmer on any day, having no prior knowledge of any beach contamination from the outfall. Further details are given in McBride (2005).

As described so far, the QMRA models are static, not having an explicit time dimension. Even though summer versus winter conditions can be examined (as in Table 8.5), each iteration of the Monte Carlo technique represents a static ('Groundhog') day. Dynamic risk assessment methodology is a recent development (Eisenberg *et al.*, 2002; Soller *et al.*, 2003, 2004), in which individuals can move from one state to another on successive days. Soller *et al.* (2003) apply this approach to a wastewater discharge (Stockton, CA) to demonstrate a low year-round risk attributable to improved treatment of the City's wastewater.

8.4   Discussion: comparing the two approaches

The general conclusion to be drawn from the above is that recreational water standards should be set using the findings of epidemiological studies, for the reasons summarized in Table 8.6, but also accompanied by QMRA studies – Eisenberg *et al.*

Table 8.5   Example results from the Christchurch City QMRA study: incremental adenovirus-related health risks per 1000 exposure events at South Brighton Beach attributable to a 2 km outfall without UV disinfection.[a]

Percentile	Summer	Winter
Minimum	0	0
90th	0	0
95th	0	1
98th	0	2
99th	0	3
99.9th	1	4
Maximum	1	6
IIR (%)[b]	0.0007	0.0151

[a] The final scheme includes a 3 km outfall.
[b] Individual infection rate, calculated as the total number of calculated infection cases divided by the total number of exposures. Ideally, it should be accompanied by a measure of uncertainly.

(2006) have noted that the two approaches are complementary and their joint use can strengthen findings about waterborne disease. Of course, for combinations of pathogens and fecal sources for which no epidemiological study has been performed, QMRA is the only available modeling tool.

Also, for recreational activity in proximity to a well-treated wastewater, QMRA is much to be preferred for the following reasons:

Table 8.6   Comparing epidemiological and QMRA approaches for setting beach criteria.

Issue	Epidemiological studies	QMRA
Represents the whole community and their swimming patterns?	Yes (but only in uncontrolled studies).	No. Dose–response data are only available for healthy adults.
Minimizes classification bias (swimmer/non-swimmer)?	Yes, but only in controlled randomized studies. (In uncontrolled trials, non-swimmers tend to be less fit and older.)	Yes.
Reflects all the pathogens present?	Yes. Because swimmers are exposed to the pathogens actually present.	No. Only includes the pathogens selected in the hazard assessment step. May ignore stirring of sediments

		releasing stored pathogens (lakes and rivers).
Reflects all the strains of a pathogen?	Yes. Because swimmers are exposed to the pathogens actually present.	No. Restricted to the few strains for which data are available[a] (usually a single strain).[b] Some important pathogens have been lacking dose–response data (especially norovirus).
Reflects the whole range of exposures?	No. Restricted to the conditions present on the epidemiological survey days.	Yes. Allows virtually unlimited exposure events.
Detects infection?	No.	Yes.
Detects illness?	Yes.	Not for all pathogens modeled.[c]

[a] These are: rotavirus, hepatitis A virus, adenovirus 4, echovirus 12, coxsackievirus, *Salmonella* (non-typhoid), *Shigella*, non-enterohemorrhagic strains of *E. coli* (except O111), *Campylobacter jejuni (Penner serotype* 27 – Black *et al.*, 1988), *Vibrio cholera, Entamoeba coli, Cryptosporidium parvum, Giardia lamblia* (Haas *et al.*, 1999).

[b] A notable exception is oocysts of *Cryptosporidium parvum*, for which a number of strains have been studied (Okhuysen *et al.*, 1999; Teunis *et al.*, 2002a, 2002b).

[c] With the use of literature data from volunteer experiments, examples can be found for three possible alternatives: an increase in the probability of illness with increasing dose (salmonellosis), a decrease with higher doses (campylobacteriosis), and a probability of illness (given infection) independent of the ingested dose (cryptosporidiosis). These alternatives may reflect different modes of interactions between pathogens and hosts (Teunis *et al.*, 1999). However, a reconsideration using new data casts doubt on the pattern previously observed for campylobacteriosis, particularly for children (Teunis *et al.*, 2005).

- The epidemiological studies have not used swimming areas in close proximity to major wastewater outfalls.

- Generally, distant wastewaters that did impact the epidemiological beaches had been chlorinated (Cabelli, 1983), which is generally a lower level of disinfection occurring in many places today.

- The associations between bacterial indicators and health risk identified in epidemiological studies (intestinal enterococci or *E. coli*) break down as the degree of wastewater treatment is enhanced (Simpson *et al.*, 2003).

QMRA is particularly attractive for examining the *relative* risks associated with wastewater treatment and disposal options (Westrell *et al.*, 2004). Further guidance on the use of probabilistic analysis for risk assessment is given by US EPA (1997).

References

ANZECC and ARMCANZ (2000) *Australian and New Zealand Guidelines for Fresh and Marine Water Quality 2000*. Paper No. 4, Volume 3, Table 9.2.3. Australian and New Zealand Environment and Conservation Council, Agricultural and Resource Management Council of Australia and New Zealand.

Armitage, P., and Berry, G. (1987) *Statistical Methods in Medical Research*. Blackwell, Oxford.

Ashbolt, N.J., and Bruno, M. (2003) Application and refinement of the WHO risk framework for recreational waters in Sydney, Australia. *Journal of Water and Health*, 1(3): 125–131.

Baden, D.G., Mende, T.J., Bikhazi, G., and Leung, I. (1982) Bronchoconstriction caused by Florida red tide toxins. *Toxicon*, 20: 929.

Black, R.E., Levine, M.M., Clements, M.L., Hughes, T.P., and Blaser, M.J. (1988) Experimental *Campylobacter jejuni* infection in humans. *Journal of Infectious Diseases*, 157: 472–479.

Cabelli, V.J. (1983) Health effects criteria for marine recreational waters, EPA-600/1-80-003. US Environmental Protection Agency, Cincinnati, OH.

Cabelli, V.J. (1989) Swimming-associated illness and recreational water quality criteria.*Water Science and Technology*, 21(2): 13–21.

Cabelli, V.J., Levin, M.A., Dufour, A.P., and McCabe, L.J. (1975) The development of criteria for recreational waters. In A.L.H. Gameson (ed.), *Waste Disposal in the Marine Environment*, pp. 63–73. Pergamon Press, Oxford.

Calderon, R.L., Mood, E.W., and Dufour, A.P. (1991) Health effects of swimmers and nonpoint sources of contaminated water. *International Journal of Environmental Health Research*, **1**: 21–31.

Carr, R. (2001) Excreta-related infections and the role of sanitation in the control of transmission. In L. Fewtrell and J. Bartram (eds), *Water Quality: Guidelines, Standards and Health*, pp. 89–113. IWA Publishing, London.

Chang, F.H., Chiswell, S.M., and Uddstrom, M.J. (2001) Occurrence and distribution of *Karenia brevisulcata* (Dinophyceae) during the 1998 toxic outbreaks on the central east coast of New Zealand. *Phycologia*, 40: 215–222.

Colford, J.M. Jr., Wade, T.J., Schiff, K.C., Wright, C.C., Griffith, J.F., Sandhu, S.K., Burns, S., Sobsey, M., Lovelace, G., and Weisberg, S.B. (2007) Water quality indicators and the risk of illness at beaches with nonpoint sources of fecal contamination. *Epidemiology*, 18(1): 27–35.

Crabtee, R.W., Cluckie, I.D., Forster, C.F., and Crockett, C.P. (1997) Waterborne adenovirus: a risk assessment. *Water Science and Technology*, 35(11–12): 1–6.

Doll, R. (1991) Sir Austin Bradford Hill and the progress of medical science. *British Medical Journal*, 305: 1521–1526.

Dufour, A.P. (1984) Health effects criteria for fresh recreational waters, EPA-600/1-84-004. Cincinnati OH: US Environmental Protection Agency.

Eisenberg, J.N.S., Brookhart, M.A., Rice, G., Brown, M., and Colford, J.M. Jr (2002) Disease transmission models for public health decision making: analysis of epidemic and endemic conditions caused by waterborne pathogens. *Environmental Health Perspectives*, 110(8): 783–790.

Eisenberg, J.N.S., Hubbard, A., Wade, T.J., Sylvester, M.D., LeChevallier, M.W., Levy, D.A., and Colford, J.M. Jr. (2006) Inferences drawn from a risk assessment compared directly with a randomized trial of a home drinking water intervention. *Environmental Health Perspectives*, 114(8): 1199–1204.

Ferley, J.P., Zmirou, D., Balducci, F., Baleux, B., Fera, P., Larbaigt, G., Jacq, E., Moissonnier, B., Blineau, A., and Boudot, J. (1989) Epidemiological significance of microbiological pollution criteria for river recreational waters. *International Journal of Epidemiology*, 18: 198–205.

Fleisher, J.M., Kay, D., Salmon, R.L., Jones, F., Wyer, M.D., and Godfree, A.F. (1996) Marine waters contaminated with domestic sewage: nonenteric illnesses associated with bather exposure in the United Kingdom. *American Journal of Public Health*, 86(9): 1228–1234.

Fleisher, J.M., Kay, D., Wyer, M.D., and Godfree, A.F. (1998) Estimates of the severity of illnesses associated with bathing in marine recreational waters contaminated with domestic sewage. *International Journal of Epidemiology*, 27: 722–726.

Gale, P. (2003) Developing risk assessments of waterborne microbial contaminations. In D. Mara and N. Horan (eds), *Handbook of Water and Wastewater Microbiology*, pp. 263–280. Academic Press, Amsterdam.

Gammie, A.J., and Wyn-Jones, A.P. (1997) Does hepatitis A pose a significant health risk to recreational water users? *Water Science and Technology*, 35(11–12): 171–177.

Greenland, S., and Robins, J.M. (1985) Estimation of a common effect parameter from sparse follow-up data. *Biometrics*, 41: 55–68.

Haas, C.N., Crockett, C.S., Rose, J.B., Gerba, C., and Fazil, A.M. (1996) Assessing the risk posed by oocysts in drinking water. *Journal of the American Water Works Association*, 88(9): 131–136.

Haas, C.N., Rose, J.B., and Gerba, C.P. (1999) *Quantitative Microbial Risk Assessment*. John Wiley & Sons, Inc., New York.

Helms, M., Vastrup, P., Gerner-Smidt, P., and Mølbak, K. (2003) Short and long term mortality associated with foodborne bacterial gastrointestinal infections: registry based study. *British Medical Journal*, 326: 357–361.

Hill, A.B. (1965) The environment and disease: Association or causation? *Proceedings of the Royal Society of Medicine*, 58: 295–300.

Jacangelo, J.G., Loughran, P., Petrik, B., Simpson, D., and McIlroy, C. (2003) Removal of enteric viruses and selected microbial indicators by UV irradiation of secondary effluent. *Water Science and Technology*, 47(9): 193–198.

Kay, D., Fleisher, J.M., Jones, F., Salmon, R.L., Wyer, M.D., Godfree, A.F., Zelenauch-Jacquotte, Z., and Shore, R. (1994) Predicting the likelihood of gastroenteritis from sea bathing: results for a randomized exposure. *Lancet*, 344: 905–909.

Kay, D., Bartram, J., Prüss, A., Ashbolt, N., Wyer, M., Fleisher, J., Fewtrell, L., Rogers, A., and Rees, G. (2004) Derivation of numerical values for the World Health Organization guidelines for recreational waters. *Water Research*, 38: 1296–1304.

Makri, A., Modarres, R., and Parkin, R (2004) Cryptosporidiosis susceptibility and risk: a case study. *Risk Analysis*, 24(1): 209–220.

Masago, Y., Katayama, H., Hashimoto, A., Hirata, T., and Ohgaki, S. (2002) Assessment of risk of infection due to *Crypotosporidium parvum* in drinking water. *Water Science and Technology*, 46(11–12): 319–324.

McBride, G.B. (2005) *Using Statistical Methods for Water Quality Management: Issues, Problems and Solutions*. John Wiley & Sons, Inc., Hoboken, NJ.

McBride, G.B., Salmond, C.E., Bandaranayake, D.R., Turner, S.J., Lewis, G.D., and Till, D.G. (1998) Health effects of marine bathing in New Zealand. *International Journal of Environmental Health Research*, 8: 173–189.

McCullagh, P., and Nelder, J.A. (1989) *Generalized Linear Models*. 2nd edn Chapman & Hall, London.

Medical Research Council (1959) *Sewage Contamination of Bathing Beaches in England and Wales*. Medical Research Council Memorandum No. 37.

Mena, K.D., Gerba, C.P., Haas, C.N., and Rose, J.B (2003) Risk assessment of waterborne coxsackievirus. *Journal of the American Waterworks Association*, 95(7): 122–131.

Ministry for the Environment and Ministry of Health (2003) *Microbiological Water Quality Guidelines for Marine and Freshwater Recreational Areas*. Ministry for the Environment and Ministry of Health, Wellington, New Zealand (http://www.mfe.govt.nz).

Moore, B. (1975) The case against microbial standards at bathing beaches. In A.L.H. Gameson (ed.), *Waste Disposal in the Marine Environment*, pp. 103–109, Pergamon Press, Oxford.

Nachamkin, I., Allos, B.M., and Ho, T.W. (2000) *Campylobacter jejuni* infection and the association with Guillain–Barré syndrome. In I. Nachamkin and M.J. Baser (eds), *Campylobacter*, 2nd edn, pp. 155–175. American Society for Microbiology, Washington, DC.

National Health and Medical Research Council (2005) *Guidelines for Managing Risks in Recreational Water*. National Health and Medical Research Council of Australia, Canberra, ACT.

National Technical Advisory Committee (1968) *Water Quality Criteria*. Federal Water Pollution Control Administration, Department of the Interior, Washington, DC.

Okhuysen, P.C., Chappell, C.L., Crabb, J.H., Sterling, C.R., and DuPont, H.L. (1999) Virulence of three distinct *Cryptosporidium parvum* isolates for healthy adults. *Journal of Infectious Diseases*, 180: 1275–1281.

Pike, E.B. (1994) Health effects of sea bathing (WMI 9021) – Phase III. Final report to Department of the Environment. Report DoE 3412/2. Water Research Centre, Medmenham, England.

Pouillot, R., Beaudeau, P., Denis, J.-B., and Derouin, F. (2004) A quantitative risk assessment of waterborne cryptosporidiosis in France using second-order Monte Carlo simulation. *Risk Analysis*, 24(1): 1–17.

Prüss, A. (1998) Review of epidemiological studies on health effects from exposure to recreational water. *International Journal of Epidemiology*, 27: 1–9.

Public Health Laboratory Service (1959) Sewage contamination of coastal bathing beaches in England and Wales: a bacteriological and epidemiological study. *Journal of Hygiene*, 57: 435–472.

Rose, J.B., and Sobsey, M. (1993) Quantitative risk assessment for viral contamination of shellfish and coastal waters. *Journal of Food Protection*, 56(12): 1043–1050.

Seyfried, P.L., Tobin, R.S., Brown, N.E., and Ness, P.F. (1985) A prospective study of swimming-related illness. II. Morbidity and the microbiological quality of water. *American Journal of Public Health*, 75: 1068–1076.

Simpson, D., Jacangelo, J., Loughran, P., and McIlroy, C. (2003) Investigation of potential surrogate organisms and public health risk in UV irradiated secondary effluent. *Water Science and Technology*, 47(9): 37–43.

Soller, J.A., Olivieri, A.W., Crook, J., Cooper, R.C., Tchobanoglous, G., Parkin, R.T., Spear, R.C., and Eisenberg, J.N.S. (2003) Risk-based approach to evaluate the public health benefit of additional wastewater treatment. *Environmental Science and Technology*, 37(9): 1882–1891.

Soller, J.A., Olivieri, A.W., Eisenberg, J.N.S., Sakkaji, R., and Danielson, R. (2004) *Evaluation of Microbial Risk Assessment Techniques and Applications*. Water Environment Research Foundation Report 00-PUM-3, Alexandria, VA.

Stevenson, A.H. (1953) Studies of bathing water quality and health. *American Journal of Public Health*, 43: 1071–1076.

Teunis, P.F.M., van der Heijden, O.G., van der Giessen, J.W.B., and Havelaar, A.H. (1996) *The Dose-Response Relation in Human Volunteers for Gastro-intestinal Pathogens*. Report no. 284550002. RIVM, Bilthoven, The Netherlands.

Teunis, P.F.M., Nagelkerke, J.D., and Haas, C.N. (1999) Dose response models for infectious gastroenteritis. *Risk Analysis*, 19(6): 1251–1260.

Teunis, P.F.M., Chappell, C.L., and Ockhuysen, P.C. (2002a) Cryptosporidium dose response studies: variation between isolates. *Risk Analysis*, 22(1): 175–183.

Teunis, P.F.M., Chappell, C.L., and Ockhuysen, P.C. (2002b) Cryptosporidium dose response studies: variation between hosts. *Risk Analysis*, 22(3): 475–485.

Teunis, P., van den Brandhof, W., Nauta, M., Wagenaar, J., van den Kerkhof, H., and van Pelt, W. (2005) A reconsideration of the *Campylobacter* dose-response relation. *Epidemiology and Infection*, 133: 583–592.

Thompson, S.S., Jackson, J.L., Suva-Castillo, M., Yanko, W.A., El Jack, Z., Kuo, J., Chen, C.L., Williams, F.P., and Schnurr, D.P. (2003) Detection of infectious human adenoviruses in tertiary-treated and ultraviolet disinfected wastewater. *Water Environmental Research*, 75: 163–170.

US Environmental Protection Agency (1986) *Quality Criteria for Water*. Report EPA 600/4-85-076. US EPA, Washington DC.

US Environmental Protection Agency (1997) *Guiding Principles for Monte Carlo Analysis*. Report EPA/630/R-97/001 (http://www.epa.gov/ncea/raf/montecar.pdf).

Wade, T.J., Pai, N., Eisenberg, J.N.S., and Colford, J.M. Jr. (2003) Do U.S. Environmental Protection Agency water quality guidelines for recreational waters prevent gastrointestinal illness? A systematic review and meta-analysis. *Environmental Health Perspectives*, 111(8): 1102–1109.

Wade, T.J., Calderon, R.L., Sams, E., Beach, M., Brenner, K.P., Williams, A.H., and Dufour, A.P. (2006) Rapidly measured indicators of recreational water quality are predictive of swimming-associated gastrointestinal illness. *Environmental Health Perspectives*, 114(1): 24–28.

Westrell, T., Schönning, C., Stenström, T.A., and Ashbolt, N.J. (2004) QMRA (quantitative microbial risk assessment) and HACCP (hazard analysis and critical control points) for management of pathogens in wastewater and sewage sludge treatment and reuse. *Water Science and Technology*, 50(2): 23–30.

World Health Organization (2003) *Guidelines for Safe Recreational Water Environments. Volume 1, Coastal and Fresh Waters*. World Health Organization, Geneva.

Wyer, M.D., Kay, D., Fleisher, J.M., Salmon, R.I., Jones, F., Godfree, A.F., Jackson, G., and Rogers, A. (1999) An experimental health-related classification for marine waters. *Water Research*, 33(3): 715–722. Correction: 33(10): 2467.

Wymer, L.J., and Dufour, A.P. (2002) A model for estimating the incidence of swimming-related gastrointestinal illness as a function of water quality indicators. *Environmetrics*, 13: 669–678.

Wymer, L.J., Dufour, A.P., Calderon, R.L., Wade, T.J., and Beach, M. (2005) Comment on 'Derivation of numerical values for the World Health Organization guidelines for recreational waters'. *Water Research*, 39: 2774–2777.

9

A plausible model to explain concentration – response relationships in randomized controlled trials assessing infectious disease risks from exposure to recreational waters

Albrecht Wiedenmann

Infectious Disease Control and Environmental Hygiene, Public Health Office, Administrative District Esslingen, Germany

9.1 Introduction

Current standards for the microbiological quality of recreational water are often based on mathematical functional forms that have been fitted to results from epidemiological investigations to model the increasing probability of adverse health effects observed at increasing concentrations of fecal indicator organisms (FIOs) in

Statistical Framework for Recreational Water Quality Criteria and Monitoring   Edited by Larry J. Wymer
© 2007 John Wiley & Sons, Ltd

the water (Commission of the European Communities, 2002; Wymer and Dufour, 2002; World Health Organization (WHO), 2003; Kay *et al.*, 2004). These functional forms, however, do not seem to take explicitly into account the special factors and circumstances that potentially or actually influence relationships between the concentrations of microbial indicators of fecal water pollution and infectious disease risks for bathers. At least they do not contain parameters or constants that correspond with known influence factors, and thus they do not seem to provide plausible explanations for the various results found in epidemiological investigations. The current WHO guidelines for recreational waters, for example, mainly rely on results from a randomized controlled trial performed at British seawater bathing sites (Kay *et al.* 1994, 2004; WHO, 2003). The functional form – a square root function – that has been used to model the results of this epidemiological investigation has been cut at a concentration of 158 fecal streptococci (FS) per 100 ml, because otherwise it would have predicted that approximately 90 % of the bathers would acquire symptomatic gastroenteritis (defined as vomiting or diarrhea with three or more bowel movements per day) from 10 min exposure including three or more head immersions in recreational waters with concentrations above approximately 1000 FS per 100 ml. If this attributable risk were added to the baseline risk for non-bathers the probability of acquiring gastroenteritis for bathers exposed at 1000 FS per 100 ml or higher would be 100%. This was clearly in contrast to former epidemiological findings, for example those by Cabelli (1983), who found attributable risks of gastroenteritis of less than 5% even at geometric mean concentrations of approximately 9000 FS per 100 ml at Egyptian beaches. The scientific board advising WHO therefore recommended not to extrapolate this functional form beyond the range of microbial concentrations that had been observed in this randomized controlled trial (that is, beyond 158 FS per 100 ml). Above 158 FS per 100 ml the functional form was consequently replaced by a horizontal line to calculate estimates for the attributable risk ('disease burden') at higher concentration levels. Because of the hypothetical character and the implausible results from extrapolation of the functional form used to model the results from the randomized controlled seawater trial, Wiedenmann *et al.* (2006) initially did not use any mathematical function to model their results from a similar randomized controlled trial at German freshwater bathing sites. They simply described their results by calculation of the incidence rates of disease (in %) below and above a no-observed-adverse-effect level (NOAEL), and by classifying the incidence rates in quartile and quintile ranges of microbial exposure concentrations.

In the present chapter functional forms (equations) are developed that include parameters and constants to explicitly describe various factors known to influence relations between FIOs and incidence rates of infectious diseases which are transmitted via the fecal-oral route by exposure to recreational waters. The equations are then used to model results from dose – response experiments with human pathogens in human volunteers, and to model the experimental data from the German and the British randomized controlled trials.

9.2 Basic considerations

9.2.1 The difference between a concentration and a dose

In almost all former publications describing relationships between the concentration of FIOs in recreational waters and the probability of adverse health effects the term 'dose–response' is used to describe such relationships. In fact, however, there is a fundamental difference between a 'dose' of pathogenic organisms (POs) and a concentration of POs, and there is another fundamental difference between a concentration of POs and a concentration of FIOs.

A dose of POs would simply be the intake[1] of a certain number of POs by a certain individual. An exposure concentration is only indicative of a dose if the intensity of exposure and, in the case of fecal-orally transmittable POs, the volume intake of water via the mouth, the nose, and the eyes and lacrimal ducts are known. In this case the dose is the concentration of POs in the water multiplied by the volume intake of water. Plotting risk of infection on the y-axis by the concentration of POs in a unit volume (say, per 100 ml) of water on an x-axis could, for example, be compared to plotting the risk of getting drunk on the y-axis by the concentration of alcohol (say, 5% for beer, 10% for wine, and 40% for spirits) on the x-axis. Without the knowledge of the volume intake this would not reveal a 'dose – response' relationship, because, for example, a 20 ml measure of spirits contains one third as much alcohol as half a liter of beer. The position of a concentration – response curve (CRC) on the x-axis depends on the volume intake. For example, if a volume of 10 ml were administered to all participants of an exposure experiment the position of the curve describing the associated risk would shift towards 10 times higher concentrations on the x-axis compared to a curve resulting from exposure to 100 ml. These factors may have contributed to the variability of the curves that have resulted from the various epidemiologic investigations into health risks from recreational water exposure. This suggests that the term 'concentration–response' should be used rather than 'dose–esponse' in order to clearly differentiate between these two fundamentally different situations.

9.2.2 The difference between a concentration of POs and a concentration of FIOs

Another factor which complicates the relationship between microbial indicators of water quality and adverse health effects is the fact that the water quality indicators are not the pathogens themselves. With only rare exceptions (e.g., enterohemorrhagic *Escherichia coli*), the FIOs belong to the normal bacterial flora of the guts of warm-blooded animals and human beings. They are therefore a direct measure of the degree of fecal pollution of the water, but only an indirect measure of the

---

[1] The amount of a substance that is taken into the body, regardless of whether or not it is absorbed (United States National Library of Medicine, 2006).

possible concentration of POs. The ratio between the number of POs and FIOs may vary according to the spread of infection (symptomatic or asymptomatic, acute or chronic) at the source of the fecal pollution. If the fecal pollution originates from a large human population, for example the sewage of a town with 100 000 inhabitants, it will always be associated with a certain minimal amount of POs, because a certain percentage of people in a large population will always excrete a certain number of POs which are endemic in the population. During an epidemic, however, the number of POs in the sewage may dramatically increase in relation to the number of FIOs, that is, the pathogen – indicator ratio (PIR) may increase considerably. If the fecal pollution, however, originates from an accidental release of feces of only a single healthy individual (animal or human) it may not be associated with any POs at all. On the other hand, if it originates from an individual with acute symptomatic disease (e.g., acute diarrhea) the concentration of excreted POs (for example, viruses) in the released fecal material may even be higher than the concentration of FIOs.

9.2.3   'The world consists of aggregates' (F. Tiefenbrunner)

Microbiological methods used to measure the concentration of FIOs as well as POs usually determine the number of colony-forming units (CFUs), colony-forming particles, most probable numbers (MPN), or, in the case of viruses and phages, the number of plaque-forming units, focus-forming units (FFUs), plaque-forming particles (PFPs), or cell culture infectious doses for 50% of the inoculated cell or tissue cultures. The reason for using terms such as 'units' or 'particles' instead of simply using the term 'organisms' is that microorganisms often form aggregates, that is, they are attached to each other or several of them are attached to other particles such as soil and sand or organic fecal material. Viruses that multiply in epithelial cells of the guts may also be enclosed in dead cells. When these aggregates are trapped, for example on a filter membrane, which is subsequently put on a nutrient agar and incubated, each of the aggregates forms a colony of organisms that is macroscopically visible and countable. Therefore, a colony may have originated from a single organism as well as from a variable number of agglomerated organisms. With regard to the infectivity for an individual who swallows such an aggregate of POs it makes a big difference, however, whether the aggregate consists of two, ten or even a thousand organisms. The infectivity of a certain volume of water with a certain concentration of CFUs of FIOs which are, for example, associated with a certain number of PFPs of viruses, may therefore be relatively higher when the water has been contaminated with fresh untreated fecal material that consists of lots of aggregates compared to a contamination with secondary or tertiary treated sewage where the process of treatment has disrupted most of the aggregates. In epidemiologic investigations, however, the outcome variable, which is usually related to the concentration of FIOs, consists of the number of symptomatic diseases which occur due to intake of certain volumes of water. The mean dose of POs which is associated with a certain concentration of FIOs may therefore be higher when aggregates are present. If sedimentation of aggregates or disruption and dilution of aggregates in the water environment happen more quickly than the die-off of POs, this would result in a CRC which is steeper than expected, because the real dose of POs is systematically underestimated by the concentration of FIOs at high levels of water pollution. In other words, the 'real' PIR in contaminated water may vary, depending on the total amount, the distribution, and the degree of disruption of aggregates, whereas the 'detectable' PIR (i.e., detectable with microbiological methods) may not indicate this variability.

9.2.4   Different decay rates of FIOs and POs

Another factor which contributes to the variability of PIRs in fecally contaminated water is the 'age' of contamination. Due to different decay rates of FIOs and POs in the environment the PIRs may change in the course of time, once the fecal material has been released into the water. The decay rates depend on the structure (robustness) of the organisms themselves, the influence of sunlight, temperature, pH, salinity, adsorption, microbial degradation, and sedimentation, to name the most important factors. FIOs may be more resistant than POs or vice versa, and even the ratios between different FIOs, for example *E. coli* and intestinal enterococci, are variable due to the variability of these phenomena (Borrego *et al.*, 1990; Dizer *et al.*, 2005; Wiedenmann *et al.*, 2004).

9.2.5   Misclassification bias

A key factor which influences detectable concentration – response relationships in epidemiologic investigations is nondifferential misclassification bias (Fleisher, 1990; Prüss, 1998; WHO, 2003; Kay *et al.*, 2004). Misclassification bias occurs if the outcome variable (the incidence rate of symptomatic disease) is erroneously and randomly assigned to false high or false low exposure concentrations. Nondifferential misclassification bias always leads to erroneously shallow CRCs with NOAELs which are too low and maximal observed adverse levels (MaxOAELs) which are too high. The two main phenomena contributing to misclassification are the imprecise assignment of microbiological monitoring results to certain exposed individuals, and the imprecision of microbiological methods (Fleisher, 1990). An epidemiological study design that minimizes these two kinds of imprecision (randomized controlled trial design) therefore also minimizes misclassification bias. Misclassification bias is only of importance, however, if it occurs between NOAEL and MaxOAEL. Below and above these values, where the slope of the CRC is zero, misclassification would have no effect.

9.3   Components of a plausible model

According to the basic considerations described above, and taking into consideration some well-known phenomena associated with infections (such as immunity and the possibility of asymptomatic disease), a model which is supposed to explain results from epidemiological investigations into the risk of acquiring infectious diseases from bathing at increasing levels of FIOs in recreational waters should comprise the following components:

1. *a baseline risk* (*BR*), that is, the risk of acquiring the same kind of disease in an unexposed control group (non-bathers) within the same observation period. *BR* is synonymous with the incidence rate of disease among the unexposed, if 'incidence rate' is defined as the number of cases in this group divided by the total number of participants in this group within a defined period of time.

2. *an attributable risk* (*AR*), that is, the risk which can be attributed to or is due to exposure. *AR* is sometimes also called 'excess risk'. *AR* may split into two components: a water-related attributable risk (AR_{wr}), and a dose-related (AR_{dr}) or concentration-related one (AR_{cr}), depending on whether a dose or a concentration is used as the determining variable. AR_{wr} is the risk which is attributable to the exposure to water but is independent of the dose or concentration of the POs or FIOs under observation. Swallowing water with a high salinity, for example, may cause diarrhea symptoms due to the osmotic effect. This effect, however, would not depend on the degree of fecal pollution of the water. It would simply overlay the effect which depends on the concentration of FIOs, that is, it would add up to the concentration – response effect. A comparable phenomenon has been described for the concentration of aeromonads and skin ailments from freshwater exposure (Wiedenmann *et al.*, 2004). The dose- or concentration-related attributable risk is the risk which is expected to increase, depending on the intake of an increasing number of POs (dose) or the degree of fecal pollution of the water (concentration of FIOs).

3. *a dose- or concentration-related maximal attributable risk* ($AR_{dr,max}$ or $AR_{cr,max}$), that is, a risk level which is reached when all susceptible individuals have been infected. *BR* plus AR_{wr} plus $AR_{dr,max}$ or $AR_{cr,max}$ make up the maximum risk level (*MR*), and *MR* plus the portion of unsusceptible (that is, either immune or asymptomatic) participants of a study cohort together account for all of the participants. The portion of unsusceptible members of the cohort can also be characterized as 'non-responders' or as being 'never at risk' (*NR*).

4. *a functional form* $f(x)_{DRR}$ describing the dose – response relationship (DRR), that is, the increase in risk dependent on the dose intake which consists of an increasing number of x POs. Alternatively, $f(x)_{DRR}$ may have to be replaced by a functional form $f(x)_{CRR}$ describing a concentration – response relationship (CRR), that is an increase in risk dependent on the exposure concentration x of either pathogenic organisms or fecal indicator organisms in the water.

5. *a pathogen – indicator ratio*, that is, the number of POs divided by the number of FIOs, if the concentration of FIOs is used as the determining variable.

6. *an estimate of the accidental volume intake of water* (v_{in}) of an average bather during exposure. If the dependent variable x in a DRR is a concentration of POs (CRR), and if this concentration is given in 100 ml units ($x_{PO/100\,ml}$), v_{in} must be given in 100 ml units as well ($v_{in\_100\,ml}$) to allow the calculation of the dose intake by multiplication of v_{in} with the concentration of POs.

7. *an estimate of the probability of infection, P(X = 1) or p(1), associated with the intake of a single PO*, if $f(x)_{DRR}$ is based on a binomial distribution, and the corresponding *cumulative density function* (CDF) of the form $F(x) = P(X \leq x)$, where $P(X \leq x)$ is the cumulative probability of acquiring an infection from the ingestion of x POs.

9.4 Development of model equations

Under the premises described above the following equations apply:

$$MR = 1 - NR, \tag{9.1}$$

$$MR = BR + AR, \tag{9.2}$$

$$AR = AR_{wr} + AR_{dr}. \tag{9.3}$$

From (9.2) and (9.3) it follows that

$$MR = BR + AR_{wr} + AR_{dr,max}. \tag{9.4}$$

Substituting (9.4) into (9.1) and rearranging gives:

$$BR + AR_{wr} + AR_{dr,max} + NR = 1. \tag{9.5}$$

In a dose – response model where the determining variable x is the dose, that is, the intake of a certain number of pathogens, and the dependent variable y is the risk of symptomatic infection (or the incidence rate of disease) in an exposed cohort, y can be expressed as follows:

$$y = BR + AR_{wr} + AR_{dr,max} \times f(x)_{DRR}. \tag{9.6}$$

If x is not the dose, that is, the intake of a certain number of POs, but the concentration of POs per 100 ml of water, then the dose is the concentration $x_{PO/100\,ml}$

multiplied by the average volume intake of water in 100 ml units (v_{in_100ml}). In a concentration – response model the dependent variable y is therefore:

$$y = BR + AR_{wr} + AR_{dr,max} \times f(x_{PO/100\,ml} \times v_{in\_100\,ml}). \qquad (9.7)$$

If x is not the concentration of POs, but the concentration of FIOs, then the dose is the concentration of FIOs per 100 ml multiplied by the PIR multiplied by the volume intake of water in 100 mL units:

$$y = BR + AR_{wr} + AR_{dr,max} \times f(x_{FIO/100ml} \times PIR \times v_{in\_100\,ml}). \qquad (9.8)$$

If $AR_{wr} = 0$ (9.8) can be simplified as follows:

$$y = BR + AR_{dr,max} \times f(x_{FIO/100\,ml} \times PIR \times v_{in\_100\,ml}). \qquad (9.9)$$

The most plausible assumption to explain DRRs of POs is that each individual pathogen has a certain mean probability $P(X = 1)$ or $p(1)$ of causing an infection once it has invaded a susceptible individual. Because viruses and bacteria, and even protozoan parasites such *Cryptosporidium* and *Giardia*, do not depend on any kind of interaction with each other to be able to start multiplying in the host, it is also plausible that the cumulative probability of infection $P(X \leq x)$ resulting from ingestion of x POs can be described by the CDF of a binomial distribution with

$$F(x) = P(X \leq x) = 1 - [1 - p(1)]^x. \qquad (9.10)$$

If $f(x)_{DRR}$ in (9.9) is replaced by (9.10), the concentration – response model can be expressed as follows:

$$y = BR + AR_{dr,max} \times \{1 - [1 - p(1)]^z\}, \qquad z = x_{FIO/100\,ml} \times PIR \times v_{in\_100\,ml}, \qquad (9.11)$$

or

$$y = BR + (MR - BR) \times \{1 - [1 - p(1)]^z\}, \quad \text{where } z = x_{FIO/100\,ml} \times PIR \times v_{in\_100\,ml}. \qquad (9.12)$$

It is virtually impossible to find a true and plausible mathematical compensation for the possible changes in the PIR, which may occur in the environment due to different decay rates, disruption of aggregates, sedimentation, etc. as described above. As an 'expedient' the following equation, which is based on experimental

observations of the relationship between concentrations of *E. coli* and intestinal enterococci in fresh and marine water (Borrego *et al.*, 1990; WHO 2003; Wiedenmann
et al., 2004; Dizer *et al.*, 2005), is suggested:

$$\log_{10}(PO/100\,ml) = a + b \times \log_{10}(FIO/100\,ml), \quad (9.13)$$

where a is the intercept and b is the slope of the regression equation. This equation
can be resolved for PO/100 ml:

$$PO/100\,ml = 10^q, \qquad \text{where } q = a + b \times \log_{10}(FIO/100\,ml). \quad (9.14)$$

The variation of the PIR, that is, the concentration of POs divided by the concentration of FIOs, depending on the concentration of FIOs can consequently be described
in the following way:

$$PIR = 10^q/(FIO/100\,ml). \quad (9.15)$$

Substituting (9.15) in (9.12) results in

$$y = BR + (MR - BR) \times \{1 - [1 - p(1)]^w\}, \quad (9.16)$$

where

$$w = x_{FIO/100\,ml} \times 10^q/x_{FIO/100\,ml} \times v_{in\_100\,ml} = 10^q \times v_{in\_100\,ml}$$

and

$$q = a + b \times \log_{10}(x_{FIO/100\,ml}).$$

Equation (9.16) is the final functional form which can be used for modeling
concentration–response relationships between the concentration of FIOs in 100 ml
units of water ($x_{FIO/100\,ml}$) and incidence rates of symptomatic infections observed
in a study cohort within a defined period of time after exposure to recreational
waters, provided there is no additional independent AR_{wr}.[2]

---

[2] In a spreadsheet program such as MS Excel® equation (9.16) can be programmed as follows:

$$y = BR + (MR - BR)^*(1 - (1 - p) \wedge (x^*(10 \wedge (a + b^*LOG10(x))/x)^*v))$$

or

$$y = BR + (MR - BR)^*(1 - (1 - p) \wedge (10 \wedge (a + b^*LOG10(x))^*v)),$$

where $x = x_{FIO/100\,ml}$. In addition, each of the model parameters BR, MR, $p = p(1)$, $v = v_{in\_100\,ml}$, a and b has to be
defined (named) and specified in a separate cell.

9.5   Application of the model equations

In order to test the applicability and plausibility of the complete model, and to determine plausible estimates for $p(1)$, equation (9.10) was fit to published results from experimental studies on the infectivity of various POs in human volunteers, and equations (9.12) and (9.16) were fit to results from randomized exposure to fecally polluted fresh- and seawater. The best fit was defined as the curve yielding the minimum sum of squared errors (SSE). The data points (number of cases divided by the number of individuals who had received a certain number of POs, or who were exposed at certain levels of FIOs) where weighted by the number of individuals in each of the various exposure categories. The $p(1)$ and/or other model parameters were varied until the minimum SSE was achieved. This was done using the Solver tool in Microsoft Excel 97. The effects of variation of certain model parameters were also demonstrated by sample calculations.

9.5.1   The effect of different cohort susceptibility and different infectivity

Figure 9.1 demonstrates the quite perfect congruence of model (9.10) with the experimental data on *Cryptosporidium* oocyst exposure of healthy adult volunteers. Figure 9.2 shows the result of modeling the published DRR for *Giardia lamblia*. The models predict a $p(1)$ of 0.004 for *Cryptosporidium* oocysts and a $p(1)$ of 0.01 for *Giardia* cysts. Tables 9.1 and 9.2 list the raw data used to draw Figures 9.1 and 9.2. In addition to the number of infected individuals (defined by multiplication of the protozoan parasites and shedding of oocysts/cysts with the feces), Tables 9.1

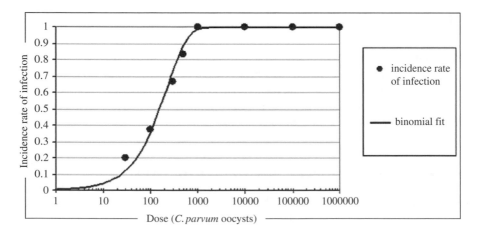

Figure 9.1   DRR for *Cryptosporidium parvum* according to DuPont *et al.* (1995), modeled using equation (9.10).

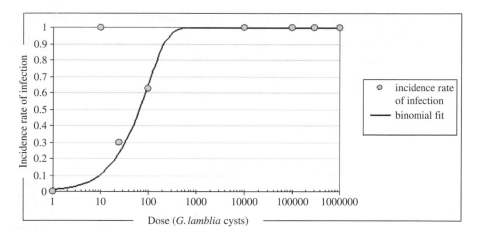

Figure 9.2   DRR for *Giardia lamblia* according to Rendtorff (1954) and Rendtorff and Holt (1954), modeled using equation (9.10).

Table 9.1   DRR for *Cryptosporidium parvum* according to DuPont et al. (1995).

Dose (oocysts)	No. of volunteers	Infected	%	95% CI [a]	Ill	%	95% CI
30	5	1	20	0.5–72.6	0	0	0–52.2
100	8	3	37.5	8.5–75.5	3	37.5	8.5–75.5
300	3	2	66.7	9.4–99.2	0	0	0–70.8
500	6	5	83.3	35.9–99.6	3	50	11.8–88.2
1 000	2	2	100	15.8–100	1	50	1.3–98.7
10 000	3	3	100	29.2–100	3	100	29.2–100
100 000	1	1	100	—	0	0	—
1 000 000	1	1	100	—	1	100	—

[a] 95% confidence interval of the binomial distribution.

and 9.2 also list the number of individuals with symptomatic disease (ill individuals) in each of the exposure categories.

The comparison of the number of infected individuals with the number of symptomatic individuals clearly demonstrates the effect of immunity or insusceptibility in a study cohort. For example, none of the individuals (adult prisoners) in the *Giardia* experiments suffered from symptoms, although large numbers of cysts were ingested. Of course, this does not mean that *Giardia* cysts generally cannot cause symptomatic disease. From epidemiological evidence we know that small numbers of cysts can definitely cause severe symptoms in susceptible individuals. It only means that this special cohort was completely resistant ($NR = 100\%$) in terms of

Table 9.2   DRR for *Giardia lamblia* according to Rendtorff (1954) and Rendtorff and Holt (1954).

Dose (cysts)	Number of volunteers	Infected	%	95% CI	Ill	%	95% CI
1	5	0	0	0–52.2	0	0	0–52.2
10	2	2	100	15.8–100	0	0	0–84.2
25	20	6	30	11.9–54.3	0	0	0–16.8
100	24	15	62.5	40.6–81.2	n. r.[a]	—	—
10 000	3	3	100	29.2–100	0	0	0–70.8
100 000	3	3	100	29.2–100	0	0	0–70.8
300 000	3	3	100	29.2–100	0	0	0–70.8
1 000 000	2	2	100	15.8–100	0	0	0–84.2

[a] Not reported.

symptomatic disease. In the *Cryptosporidium* experiments, however, some of the volunteers did suffer from gastroenteritis symptoms. To determine the most probable value for the percentage of resistant individuals in this cohort the following equation was used:

$$P(X \le x) = MR \times \{1 - [1 - p(1)]^x\}. \qquad (9.17)$$

As $p(1)$ is already known (Figure 9.1), the most probable estimate for MR (and thus NR) was easy to determine by simply varying MR and minimizing the SSE. This procedure yielded an MR of approximately 65 % and thus an NR of approximately 35 % in this cohort (Figure 9.3).

9.5.2   Effects of immunosuppression

Tarazona *et al.* (1998) investigated the effects of immunosuppression by experimentally infecting immunocompetent and immunosuppressed mice with *Cryptosporidium* oocysts. Immunosuppressed mice had a shorter prepatent period, remained infected longer and shed more oocysts than immunocompetent mice. Immunosuppression also produced high mortality rates, whereas none of the immunocompetent mice died from infection. Differences were also observed in mice infected at either 3 or 4 weeks of age. The prepatent period was shorter, and the patent period was longer in younger mice, showing that age at time of infection can modify the oocyst shedding profile. Unfortunately the authors did not attempt to produce a dose – response curve. They only tested three very high doses (10^4, 10^5, 10^6 oocysts) which proved infective for all mice. It is interesting, however, that there were no statistical differences in either prepatent or patent periods, or in

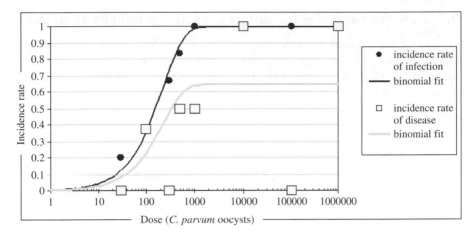

Figure 9.3 DRRs for *Cryptosporidium parvum* according to DuPont *et al.* (1995): infection versus symptomatic disease.

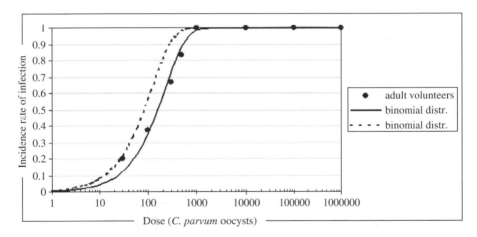

Figure 9.4 DRR for *Cryptosporidium parvum* in healthy adult human volunteers (DuPont *et al.*, 1995), and the potential effect of immunosuppression (dashed line).

the oocyst shedding profiles for these high doses. This finding provides evidence for the assumption that there is no change in effects for the uptake of doses beyond the ID_{100}, when the dose – response curve goes into a steady-state plateau. In terms of potential effects of immunosuppression on the shape of a dose–response curve, it is most likely that immunosuppression would simply cause a shift of the whole curve to the left (Figure 9.4). In terms of symptomatic disease, it can be assumed that in an immunosuppressed cohort MR would tend to reach 100%.

9.5.3   More examples for differences in the infectivity of POs

Figures 9.5 and 9.6 show the results of modeling experimental dose – response data in human volunteers for *Shigella flexneri* (Kotloff *et al.*, 1995a, 1995b), and human rotavirus (Ward *et al.*, 1986). Model (9.17) suggests a $p(1)$ of 0.002 for *Shigella flexneri*, and a $p(1)$ of 0.17 for human rotavirus.

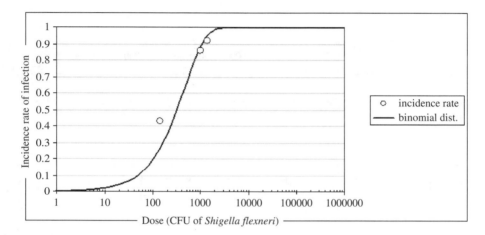

Figure 9.5   DRR for *Shigella flexneri* according to Kotloff *et al.* (1995a, 1995b), modeled using equation (9.17).

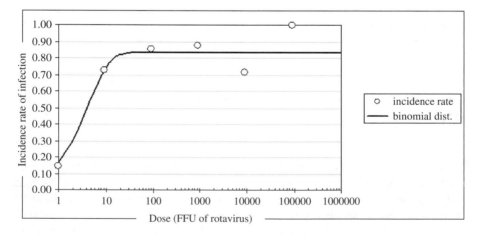

Figure 9.6   DRR for rotavirus according to Ward *et al.* (1986), modeled using equation (9.17).

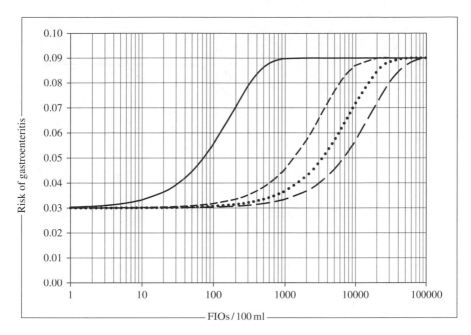

Figure 9.7   CRRs according to equation (9.12) with variable $p(1)$. Parameter settings: $BR = 0.03$; $MR = 0.09$; PIR $= 0.1$; $v_{\text{in\_100\,ml}} = 0.3$. $p(1) = 0.17$ (rotavirus, solid line), 0.01 (*Giardia*, short-dashed line), 0.004 (*Cryptosporidium*, dotted line) or 0.002 (*Shigella*, long-dashed line).

9.5.4   The theoretical effect of different $p(1)$ values

The theoretical effect of different $p(1)$ values on the shape of a CRC for FIOs is demonstrated in Figure 9.7 using equation (9.12).

9.5.5   The theoretical effect of different PIRs

The theoretical effect of different PIRs on the shape of a CRC for FIOs is demonstrated in Figure 9.8 using equation (9.12).

9.5.6   The theoretical effect of differences in the volume intake of water

The theoretical effect of differences in the volume intake of water on the shape of a CRC for FIOs is demonstrated in Figure 9.9 using equation (9.12).

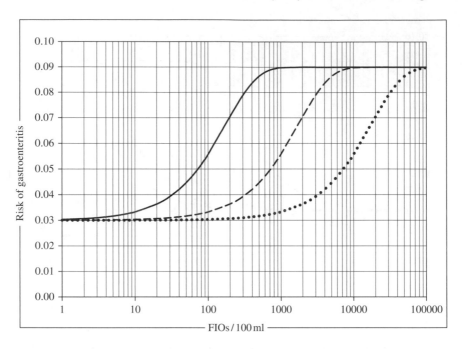

Figure 9.8   CRRs according to equation (9.12) with variable PIRs. Parameter settings: $BR = 0.03$; $MR = 0.09$; $p(1) = 0.17$; $v_{in\_100ml} = 0.3$. PIR $= 0.1$ (solid line), 0.01 (dashed line) or 0.001 (dotted line).

9.5.7   Simultaneous variability of multiple determining parameters

A comparison of Figures 9.7–9.9 demonstrates that the variation of $p(1)$, PIR or $v_{in\_100\,ml}$ would modulate the shape of the CRC in a similar way: their variation determines the position of the curve on the x-axis. Consequently, these parameters can either enhance or compensate each other. For example, a PIR $= 0.1$ and a $v_{in\_100\,ml} = 0.01$ will reveal the same CRC as a PIR $= 0.01$ and a $v_{in\_100\,ml} = 0.1$. It is therefore impossible to find a definitive solution for all three parameters at the same time by modeling a given set of epidemiological data. It is, however, possible to get an estimate for one parameter if the other two parameters are fixed. For example, if the mean volume intake of water is known from other studies, and it can be assumed that the disease was caused by a virus with a known $p(1)$, it is possible to find a most probable estimate for the PIR. On the other hand, if the model was used to test the hypothesis that gastrointestinal disease in a certain epidemiologic study was primarily caused by protozoan parasites with a low $p(1)$, and the model would reveal an implausibly high PIR or $v_{in\_100\,ml}$, it could be concluded that the hypothesis was probably wrong, and it was more plausible to assume that the causative agent was a virus.

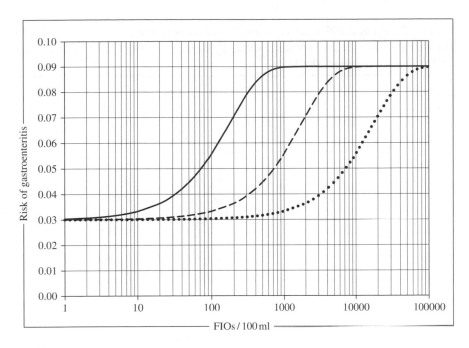

Figure 9.9 CRRs according to equation (9.12) with variable v_{in_100ml}. Parameter settings: $BR = 0.03$; $MR = 0.09$; $p(1) = 0.17$; PIR = 0.1. $v_{in\_100ml} = 0.3$ (solid line), 0.03 (dashed line) or 0.003 (dotted line).

9.5.8 Modeling results from randomized controlled trials

Tables 9.3–9.5 list results from randomized controlled exposure to freshwater (Wiedenmann *et al.*, 2006) and seawater (Kay *et al.*, 1994). In both studies the bathing duration was 10 min, and the participants were asked to immerse their head at least three times. In the freshwater trial the incidence rates of GE (gastroenteritis, number of cases / number of exposed participants) within one week after exposure were classified into quartile and quintile categories of *E. coli* and intestinal enterococci concentrations. In the seawater trial, results were classified into quartile and 20-unit categories of fecal streptococci concentrations.

In Figures 9.10, 9.12 and 9.14 equation (9.12) was fit to the geometric range means of these categories. In Figures 9.11, 9.13 and 9.15 equation (9.16) was fitted to the same set of data for comparison. To calculate the SSEs each of the data points was weighted by the number of participants in each of the categories. The non-bathers' risk was used as the *BR* in all calculations. The mean volume intake of water was assumed to be 30 ml ($v_{in\_100ml} = 0.3$), and the $p(1)$ was assumed to be 0.17 (equivalent to the $p(1)$ of rotavirus). Thus the model could be used to predict the *MR* and the PIR in equation (9.12) or the *MR* and the variation of the PIR (that is, parameters *a* and *b*) in equation (9.16). For *E. coli* concentrations in freshwater the

Table 9.3   Rates of GE within one week after randomized controlled exposure to quartile and quintile concentration categories of E. coli in freshwater according to Wiedenmann et al. (2006).

Exposure range (FIOs/100 ml)[a]	Weight (number of bathers)	Geometric range mean (FIOs/100 ml)	GE[b] rate (n cases/n bathers)
5–72	207	18	0.019
72–181	212	114	0.052
181–379	211	262	0.066
379–4600	208	1320	0.082
5–61	166	17	0.018
61–116	168	84	0.036
116–245	170	168	0.059
245–445	166	330	0.072
445–4600	168	1431	0.089
unexposed	921		0.028 [0.019, 0.041][c]

[a] Concentration determined as most probable numbers per 100 ml; figures rounded to zero decimals.
[b] Definition of GE: (diarrhea) or (vomiting) or (nausea and fever) or (indigestion and fever); combination of symptoms with fever did not occur.
[c] 95% confidence interval.

Table 9.4   Rates of GE within one week after randomized controlled exposure to quartile and quintile concentration categories of intestinal enterococci in freshwater according to Wiedenmann et al. (2006).[a]

Exposure range (FIOs/100 ml)	Weight (number of bathers)	Geometric range mean (FIOs/100 ml)	GE rate (n cases/n bathers)
3–14	203	6	0.024
14–53	203	27	0.042
53–101	196	73	0.067
101–1190	190	347	0.087
3–12	167	6	0.018
12–27	167	18	0.03
27–68	169	43	0.089
68–114	168	88	0.042
114–1190	167	368	0.096
unexposed	921		0.028 [0.019, 0.041]

[a] GE, exposure range and confidence interval as in Table 9.3.

Table 9.5 Rates of GE within three weeks after randomized controlled exposure to quartile and 20-unit concentration categories of presumptive fecal streptococci in seawater according to Kay *et al.* (1994).

Exposure range (FIOs/100 ml)[a]	Weight (number of bathers)	Geometric range mean (FIOs/100 ml)	GE[b] rate (*n* cases/*n* bathers)
0–13	159	4	0.107
14–26	109	19	0.11
27–49	121	36	0.14
50–158	118	89	0.246
0–19	184	4	0.109
20–39	161	28	0.106
40–59	82	49	0.183
60–79	57	69	0.281
80–158	23	112	0.304
unexposed			0.097 [0.07, 0.12][c]

[a] Concentration determined as colony forming units per 100 ml.
[b] Definition of GE: ($\geq$ 3 loose bowel movements in 24 h) or (vomiting) or (nausea and fever) or (indigestion and fever).
[c] 95% confidence interval.

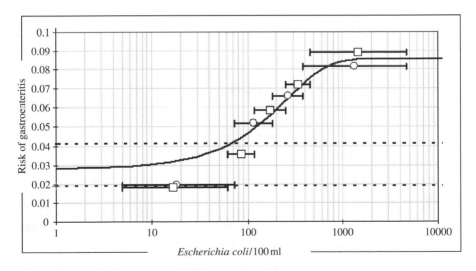

Figure 9.10 Modeling the results from randomized controlled exposure to freshwater (Table 9.3) by minimizing the SSE for equation (9.12). Parameter settings: BR (non-bathers) = 0.028; $p(1)$ = 0.17; v_{in_100ml} = 0.3. Parameter predictions: $MR = 0.086$; $PIR = 0.070$. $SSE = 15.96$.

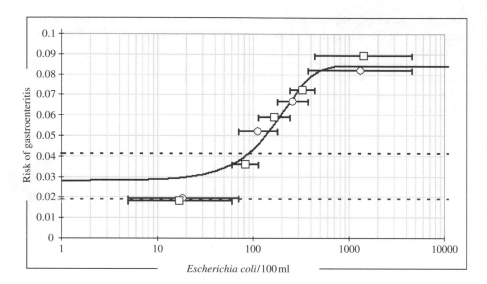

Figure 9.11   Modeling the results from randomized controlled exposure to freshwater (Table 9.3) by minimizing the SSE for equation (9.16). Parameter settings: BR (non-bathers) $= 0.028$; $p(1) = 0.17$; $v_{in\_100ml} = 0.3$. Parameter predictions: $MR = 0.084$; $a = -2.17$; $b = 1.46$. SSE $= 11.86$.

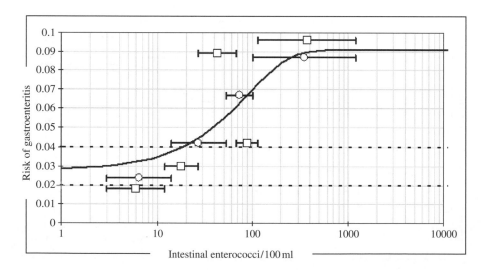

Figure 9.12   Modeling the results from randomized controlled exposure to freshwater (Table 9.4; FIO $=$ intestinal enterococci) by minimizing the SSE for equation (9.12). Parameter settings: BR (non-bathers) $= 0.028$; $p(1) = 0.17$; $v_{in\_100ml} = 0.3$. Parameter predictions: $MR = 0.091$; PIR $= 0.20$.SSE $= 70.48$.

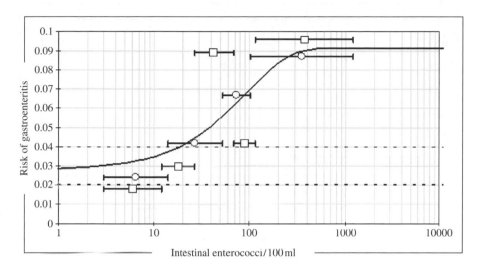

Figure 9.13 Modeling the results from randomized controlled exposure to fresh-water (Table 9.4; FIO = intestinal enterococci) by minimizing the SSE for equation (9.16). Parameter settings: BR (non-bathers) = 0.028; $p(1) = 0.17$; $v_{in\_100ml} = 0.3$. Parameter predictions: $MR = 0.091$; $a = -0.67$; $b = 0.98$. SSE = 70.46.

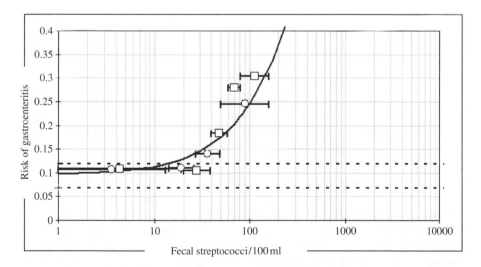

Figure 9.14 Modeling the results from randomized controlled exposure to seawater (Table 9.4; FIO = fecal streptococci) by minimizing the SSE for equation (9.12). Parameter settings: BR (non-bathers) = 0.097; $p(1) = 0.17$; $v_{in\_100ml} = 0.3$. Parameter predictions: $MR = 1.00$; PIR = 0.033. SSE = 65.19.

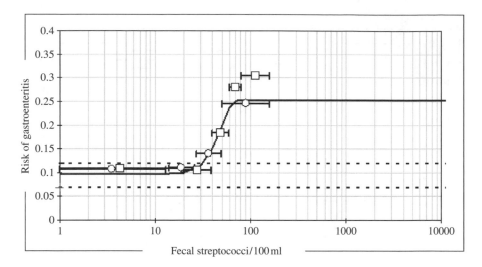

Figure 9.15 Modeling the results from randomized controlled exposure to seawater (Table 9.4; FIO = fecal streptococci) by minimizing the SSE for equation (9.16). Parameter settings: *BR* (non-bathers) = 0.097; $p(1) = 0.17$; $v_{in\_100ml} = 0.3$. Parameter predictions: $MR = 0.25$; $a = -5.92$; $b = 4.23$. SSE = 15.19.

model that assumes a constant PIR (9.12) and the model that assumes a variable PIR (9.16) revealed slightly different curves, and the model with variable PIR revealed a lower SSE, that is, a better fit, whereas there was no difference between the two curves for intestinal enterococci. This result is in accordance with the fact that in the same study and in other studies (Dizer *et al.*, 2005) the relation between the concentrations of *E. coli* and intestinal enterococci was not 1, but could be described by a relationship of the form $y = a + b \times \log_{10}$(concentration of *x*), with higher *E. coli*/enterococci relations at high concentrations than at low concentrations. This is in accordance with the well-known fact that *E. coli* is less robust in fresh- and seawater environments than intestinal enterococci. It is also in accordance with the fact that the survival of intestinal enterococci in freshwater closely resembles the survival of somatic coliphages (Borrego *et al.*, 1990), and phages may be considered to be the best indicator for the survival of pathogenic viruses. Viruses were the most likely agents that have caused the cases of gastroenteritis in these studies, because only viruses have a $p(1)$ that is big enough to explain the concentration response without the assumption of an implausibly high mean volume intake of water or an implausibly high PIR.

If equation (9.12) with the constant PIR is fit to the results from seawater exposure (Figure 9.14) the model reveals an SSE that is much higher (by a factor in excess of 4) than for the model using equation (9.16), and it reveals the implausible result that 100% of the cohort were susceptible. The model using equation (9.16), however, reveals a much better fit (Figure 9.15). It predicts an *MR* of 0.25, that is, an *AR* of 0.15, and a relative risk of 2.5. It also predicts that the PIR is higher at high concentrations of FIOs and lower at low concentrations of FIOs. If the assumption is true, that viruses in saltwater have a decay rate that is similar to the decay rate of fecal streptococci, a plausible explanation for this phenomenon might be that sewage treatment in the late 1980s and early 1990s in the UK was less advanced than in 2000 and 2001 when the freshwater studies were performed in Germany, and for this reason high concentrations of FIOs were associated with relatively higher amounts of fecal aggregates that harbored relatively higher numbers of viruses.

9.6 Conclusions

The mathematical model described here can provide plausible explanations for most of the differences between results from epidemiological investigations on health effects from exposure to recreational waters. It is especially valuable to demonstrate that it would be inappropriate to use results from only one study with certain determining parameters (for example, a certain *BR*, *MR*, and PIR) as the only basis for setting recreational water standards. In their publication describing the results of randomized controlled exposure to fresh recreational water in Germany (Wiedenmann *et al.*, 2006) the authors demonstrated that if recreational water standards are based mainly on the highly variable *MR*s, which can be observed in different studies even with a similar study design, 'acceptable' levels (for example, guide values) of FIOs in recreational waters may differ by a factor in excess of 10, depending on nothing other than different cohort susceptibilities and different definitions of gastroenteritis (more or less severe). They also stressed that standards are intended to protect the health of those consumers who are not already immune to the pathogens which may be associated with fecal indicator organisms (for example, tourists). The model described by equation (9.16) might be used to harmonize results from different studies, by adjusting them for similar *BR*s, *MR*s, PIRs and volume intakes of water, and might thus provide a useful tool to estimate health effects for various epidemiological situations as well as various ways of recreational water use that are associated, for example, with different amounts of accidental water intake.

9.7 Acknowledgments

Juan M. López-Pila inspired the development of the present model by his calculations published in Dizer *et al.* (2005).

References

Borrego, J.J., Córnax, R., Morinigo, M.A., Martinez-Manzanares, E., and Romero, P. (1990) Colipohages as an indicator of faecal pollution in water. Their survival and productive infectivity in natural aquatic environments. Water *Research*, 24: 111–116.

Cabelli, V. (1983) Health effects criteria for marine recreational waters, EPA-600/1-80-031. US Environmental Protection Agency, Cincinnati OH.

Commission of the European Communities (2002) Proposal for a directive of the European Parliament and of the Council concerning the quality of bathing water. Brussels, October 24. http://europa.eu.int/eur-lex/en/com/pdf/2002/com2002_0581 en01.pdf (accessed August 1, 2004).

Dizer, H., Wolf, S., Fischer, M., López-Pila, J.M., Röske, I., and Schmidt, R. *et al.* (2005) Die Novelle der EU-Badegewässerrichtlinie – Aspekte der Risikobewertung bei der Grenzwertsetzung. *Bundesgesundheitsblatt Gesundheitsforsch Gesundheitsschutz*, 48: 607–614.

DuPont, H.L., Chappell, C.L., Sterling, C.R., Okhuysen, P.C., Rose, J.B., and Jakubowski, W. (1995) The infectivity of *Cryptosporidium parvum* in healthy volunteers. *New England Journal of Medicine*, 332: 855–859.

Fleisher, J.M. (1990) The effects of measurement error on previously reported mathematical relationships between indicator organism density and swimming associated illness: a quantitative estimate of the resulting bias. *International Journal of Epidemiology*, 19: 1100–1106.

Havelaar, A., Blumenthal, U.J., Strauss, M., Kay, D., and Bartram, J. (2001) Guidelines: the current position. In L. Fewtrell and J. Bartram (eds) *Water Quality: Guidelines, Standards and Health*, pp. 17–42. IWA Publishing, London.

Kay, D., Fleisher, J.M., Salmon, R.L., Wyer, M.D., Godfree, A.F., Zelenauch-Jacquotte, Z., and Shore, R. (1994) Predicting likelihood of gastroenteritis from sea bathing: results from randomised exposure. *Lancet*, 344: 905–910.

Kay, D., Bartram, J., Prüss, A., Ashbolt, N., Wyer, M.D., and Fleisher, J., *et al.* (2004) Derivation of numerical values for the World Health Organization guidelines for recreational waters. *Water Research*, 38: 1296–1304.

Kotloff, K.L., Losonsky, G.A., Nataro, J.P., Wassermann, S.S., Hale, T.L., Taylor, D.N., Newland, J.W., Sadoff, J.C., Formal, S.B., and Levine, M.M. (1995a) Evaluation of the safety, immunogenicity, and efficacy in healthy adults of four doses of life oral hybrid *Escherichia coli-Shigella flexneri* 2a vaccine strain EcSf2a-2. *Vaccine*, 13: 495–502.

Kotloff, K.L., Nataro, J.P., Losonsky, G.A., Wassermann, S.S., Hale, T.L., Taylor, D.N., Sadoff, J.C., and Levine, M.M. (1995b) A modified Shigella volunteer challenge model in which the inoculum is administered with bicarbonate buffer: clinical experience and implications for Shigella infectivity. *Vaccine*, 13: 1488–1494.

Rendtorff, R.C. (1954) The experimental transmisson of human intestinal protozoan parasites. *American Journal of Hygiene* 59: 209–220.

Rendtorff, R.C., and Holt, C.J. (1954) The experimental transmission of human intestinal protozoan parasites. *American Journal of Hygiene* 60: 327–338.

Prüss, A. (1998) Review of epidemiological studies on health effects from exposure to recreational water. *International Journal of Epidemiology*, 27: 1–9.

Tarazona, R., Blewett, D.A., and Carmona, M.D. (1998) *Cryptosporidium parvum* infection in experimentally infected mice: infection dynamics and effect of immunosuppression. *Folia Parasitologica* (Praha), 45: 101–107.

United States National Library of Medicine (2006) Toxicology glossary. http://www.sis. nlm.nih.gov/enviro/glossaryi.html (accessed June 18, June 2006).

Ward, R.L., Bernstein, D.I., Young, E.C., Sherwood, J.R., Knowlton, D.R., and Schiff, G.M. (1986) Human rotavirus studies in volunteers: determination of infectious dose and serological response to infection. *Journal of Infectious Diseases* 154: 871–880.

Wiedenmann, A., Krüger, P., Gommel, S., Eissler, M., Hirlinger, M., Paul, A., Jüngst, K., Sieben, E., and Dietz, K. (2004) Epidemiological determination of disease risks from bathing. Report No. UFOPLAN 298 61 503. Federal Environmental Agency (UBA), Berlin.

Wiedenmann, A., Krüger, P., Dietz, K., López-Pila, J.M., Szewzyk, R., and Botzenhart, K. (2005) A randomized controlled trial assessing infectious disease risks from bathing in fresh recreational waters in relation to the concentration of *Escherichia coli*, intestinal enterococci, *Clostridium perfringens* and somatic coliphages. *Environmental Health Perspectives* 114: 228–236. Supplemental Material. http://ehp.niehs.nih.gov/docs/2005/8115/supplemental.pdf (accessed September 29 2005).

Wiedenmann, A., Krüger, P., Dietz, K., López-Pila, J.M., Szewzyk, R., and Botzenhart, K. (2006) A randomized controlled trial assessing infectious disease risks from bathing in fresh recreational waters in relation to the concentration of *Escherichia coli*, intestinal enterococci, *Clostridium perfringens* and somatic coliphages. *Environmental Health Perspectives*, 114: 228–236.

World Health Organization (2003) *Guidelines for Safe Recreational-Water Environments. Vol. 1: Coastal and Fresh Waters.* Geneva: World Health Organization Marketing and Dissemination.

Wymer, L.J., and Dufour, A.P. (2002) A model for estimating the incidence of swimming-related gastrointestinal illness as a function of water quality indicators. *Environmetrics*, 13: 669–678.

10

Nowcasting recreational water quality

Alexandria B. Boehm[1], Richard L. Whitman[2], Meredith B. Nevers[2], Deyi Hou[1], and Stephen B. Weisberg[3]

[1]*Department of Civil and Environmental Engineering, Stanford University, Stanford, CA, USA (AB, DH);* [2]*Lake Michigan Ecological Research Station, United States Geological Survey, Porter, IN, USA (RW, MN);* [3]*Southern California Coastal Water Research Project, Costa Mesa, CA, USA (SW)*

10.1   Introduction

Considerable resources are expended each year to measure fecal indicator bacteria (FIB) and assess whether recreational beaches are free from fecal contamination (Schiff *et al.*, 2002). However, these monitoring programs are compromised because current methods of enumerating bacteria are too slow to provide full protection from exposure to waterborne pathogens. The current United States Environmental Protection Agency (US EPA) approved methods to evaluate recreational waters require an incubation period of 18–96 hours, while several studies have shown that temporal changes in indicator bacteria levels in beach water occur on much

Statistical Framework for Recreational Water Quality Criteria and Monitoring   Edited by Larry J. Wymer
© 2007 John Wiley & Sons, Ltd

shorter time scales (Leecaster and Weisberg, 2001; Boehm *et al.*, 2002a). Thus, contaminated beaches remain open during the laboratory incubation period and are often clean by the time warnings are posted (Kim and Grant, 2004).

Advances in molecular techniques may soon provide new opportunities for measuring bacteria more rapidly (Haugland *et al.*, 2005; Noble and Weisberg, 2005). Whereas present measurement methods are based on culturing bacteria and quantifying metabolic activity, molecular methods allow direct measurement of cellular attributes such as genetic material or surface immunological properties. Removing the extended incubation step allows these methods to potentially provide results in less than 4 hours, a short enough time for managers to take action to protect public health (i.e., post a warning or close a beach) on the same day that water samples are collected.

However, even rapid methods will not solve all timing problems with the warning systems. They will not protect people who swim prior to sampling, during sample processing, and while mitigative or warning actions are being taken. They will also not provide information soon enough for beachgoers living hours from the beach to alter their day trip travel plans. In addition, rapid detection of microbial pollutants cannot address the challenge of interpreting the very small-scale temporal variability (of the order of minutes) of FIB densities in a marine surf zone (Boehm *et al.*, 2002a).

An alternative approach to providing more timely information is the use of predictive models. Predictive models use alternative information about the site, such as ocean conditions, presence of input sources, and meteorological data to identify the likelihood that bacterial concentrations will be elevated on selected dates or at selected times of those days. Such nowcasting models can be based on simple correlative relationships established empirically through prior experience, or they can be complex mechanistic models based on experimentally derived process functions.

Models can be used alone to trigger health warnings of as a means of initiating adaptive implementation of routine bacterial monitoring. Most monitoring programs in the United States sample at weekly or monthly intervals. At such low sampling frequency, monitoring needs to be focused on those time periods when concentrations are most likely to be high so that a water quality problem can be identified if one exists. Modeling can provide the means for selecting the most appropriate sampling time.

Modeling can also be used as a tool for interpreting data that are collected in routine bacterial monitoring systems. Factors such as tide, solar radiation, and wind affect bacterial concentration. Extrapolating from samples collected in the morning to conditions in the early afternoon when swimmers are most numerous will be biased by changes caused by these factors. Models can be used to integrate these factors and predict concentration for a later time in the day. Similarly, models can be effective tools in conducting trends assessment in which the models are used to deconvolve within-day variation caused by such factors from the longer-term trends of management interest.

While there are many such applications, the use of modeling in beach water quality warning systems is in the early stages of development (National Research Council, 2004). This chapter presents a summary of these developing efforts. First, we describe documented physical, chemical, and biological factors that have been demonstrated by researchers to affect bacterial concentrations at beaches and thus represent logical parameters for inclusion in a model. Then, we illustrate how various types of models can be applied to predict water quality at freshwater and marine beaches.

10.2 Factors

The fate and transport of FIB in the environment are impacted by many factors, including sunlight, rainfall, tides, waves, wind, and temperature. A handful of biological factors, such as plant wrack, birds, zooplankton, and human bathers, have also been shown to be important influences. This section provides a review of these factors and the mechanisms by which they impact pollutant levels. When constructing models, these factors should be evaluated and considered for inclusion.

10.2.1 Sunlight

For decades researchers have documented the effect of sunlight on the culturability of FIB using laboratory and *in situ* microcosm experiments and field studies of polluted marine/freshwater beaches and estuaries. At Huntington State Beach, CA, there is a distinct diurnal signal in total coliform (TC), *E. coli* (EC), and enterococci (ENT) levels in the surf zone (Boehm *et al.*, 2002a). During the dark hours of the night and early morning, FIB are typically two to ten times higher than when the sun is out. Between 12 noon and 2 p.m., densities of FIB typically fall to below detection limits of commonly used microbial assays. Historical data from Santa Cruz, CA, where water samples have been collected at various times during the day for the last 20 years also show a distinct diurnal pattern, with samples collected in the afternoon significantly lower in FIB than those collected in the early morning (Boehm, unpublished data). At 63rd Street Beach in Chicago, EC levels are significantly lower on sunny than on cloudy days (Whitman *et al.*, 2004). This effect has been observed at numerous beaches (Wymer *et al.*, 2005) and even in Antarctica near the sewage outfall at Rothera Point. During the austral summer when there are 24 hours of sunlight, fecal coliform (FC) is present only within 150 m of the outfall, while during the austral winter it extends over 500 m, beyond the region sampled (Hughes, 2003).

The mechanism whereby FIB are inactivated is a function of concentrations of exogenous or endogenous sensitizers present in the water. When excited by photons, these cause damage to DNA or other cellular components by promoting the production of free radicals in the presence of dissolved oxygen and organics. Colloidal clay can reduce light-induced damage by stabilizing excited sensitizers. Oguma *et al.*

(2001) showed that EC subjected to sunlight damage can repair themselves using intracellular repair mechanisms. These findings suggest that sunlight-mediated inactivation of FIB in laboratory and field surveys may not be permanent. Inactivation of FIB in the presence of sunlight has been estimated as first-order decay (Bellair *et al.*, 1977; McCambridge and McMeekin, 1981; Fujioka *et al.*, 1981; Fujioka and Narikawa, 1982; Sinton *et al.*, 1999; Davies-Colley *et al.*, 1994; Sinton *et al.*, 2002).

10.2.2 Rainfall

Rainfall is the only environmental factor that is presently used as the basis for proactive beach water quality health warnings. In southern California, 0.3 cm of rain typically triggers warnings to stay out of the ocean for three days (Ackerman and Weisberg, 2003). California's warnings are only advisory, but Monmouth County in New Jersey routinely closes two beaches that consistently have elevated bacterial concentrations following runoff events for 24 hours following 0.3 cm or more rain; the closure is extended to 48 hours following 7 cm or more of rain (National Research Council, 2004). Rainfall is also used as a trigger for closing shellfish beds from harvest in many states along the east coast of the USA.

Rainfall is associated with elevated bacterial indicator levels on both daily (Curriero *et al.*, 2001; Kistemann *et al.*, 2002; Schiff *et al.*, 2003) and seasonal (Lipp *et al.*, 2001; Boehm *et al.*, 2002a) time scales. Olyphant and Whitman (2004) modeled multiple environmental factors to predict bacterial concentration and found that rainfall was the most important factor positively correlated with bacterial concentration at 63rd Street Beach in Chicago.

Rainfall can impact recreational water quality via at least two different mechanisms. The first is overflow of combined stormwater/sewer systems. High rainfall can cause inflows to exceed treatment facility capacity, with excess sewage influent bypassed to receiving waters. The second mechanism is land-based flow that is directed immediately to the receiving water, which occurs where stormwater conveyance systems are distinct from the sewage treatment system. Stormwater not associated with combined sewer overflows could be tainted with FIB from environmental reservoirs, wildlife and domesticated animals, or sewage leaking from septic systems or aged sewage infrastructure. The fecal counts in land-based runoff vary greatly with the land-use type, but have been found to exceed 10^6 per 100 ml from urban areas and can be even higher near animal farms (Ackerman and Schiff, 2003).

The amount of rainfall and flow necessary to enhance bacterial concentrations varies by mechanism. For the separate systems, the amount of runoff is primarily a function of rainfall volume, the percentage of impervious surface in the watershed and the amount of antecedent rainfall. Ackerman and Weisberg (2003) found that rainfall of less than 0.3 cm in southern California did not lead to increased beach bacteria, but rainfall greater than 1.3 cm consistently led to higher bacterial levels; the effect of rainfall between 0.3 and 1.3 cm on bacterial concentration was mostly dependent on antecedent rainfall. Antecedent rainfall soaks the ground and reduces infiltration, though FIB concentration following very recent rainfall can be reduced by the previous washoff.

For combined systems, the trigger for increased bacterial concentrations is related to capacity of the local treatment facility. Once that capacity is exceeded, concentrations rise as some portion of flow is bypassed without treatment. This step function is mostly related to size of the treatment facility but can also relate to the type of conveyance systems. In some areas, even those with separate sewage and stormwater systems, rainfall infiltration into the sewage conveyance systems can lead to overflows of the piping before sewage reaches the treatment plant.

10.2.3   Tide

There are several mechanisms whereby tides might influence FIB concentrations along the shoreline. Incoming tides can dilute nearshore sources and reduce bacterial concentrations during flood conditions (Coelho *et al.*, 1999). Outgoing tides drain material from land to sea during ebb flow via surficial conduits such as streams and drains (Grant *et al.*, 2001) and through the subsurface (Urish and McKenna, 2004). Higher than average spring tides may provide a hydrologic connection between the sea and fecal sources that under average conditions would not communicate (e.g., material at the high water line or the upper reaches of the surface or subsurface tidal prism). Tidally modulated currents may also to be capable of moving FIB from a distant source to the beach (Boehm *et al.*, 2002b; Kim *et al.*, 2004).

Boehm and Weisberg (2005) found that ENT levels were significantly higher during spring tides than neap tides at 50 of the 60 marine recreational beaches they examined; for no site did they find statistically higher FIB concentrations under neap tide conditions. Spring tides may have this effect because water reaches the upper intertidal zone where FIB sources exist in the upper intertidal marine sands (Obiri-Danso and Jones, 2000; Choi *et al.*, 2003). Oshiro and Fujioka (1995) documented that bird droppings contribute to this reservoir, and Anderson *et al.* (1997) found that FIB accumulate in the plant wrack line, where they remain before resuspension during extreme tides.

Boehm and Weisberg (2005) found that the tide stage, defined as whether the tide level is falling (ebbing), or rising (flooding), had less of an impact on beach water quality than tide range at the 60 southern Californian beaches examined. However, tide stage effects have been documented at other beach locations, where FIB concentrations have generally been higher during ebb tide (Paranhos *et al.* 1998; Coelho *et al.*, 1999; Crowther *et al.*, 2001; Taggart, 2002; Kim *et al.*, 2004).

10.2.4   Waves

In a wave-dominated coastal environment such as the west coast of the USA, waves primarily control transport and mixing in the surf zone. In this context, the surf zone is the region of the coastal ocean from where the waves begin to break to the shoreline, and it is where most of the contact between bathers and marine recreational waters occurs.

The surf zone is typically assumed to be well mixed across its width due to the wave-driven turbulence. An alongshore current is induced in the surf zone by the breaking waves. Mathematically, the speed of the alongshore, wave-induced current (q_l) can be estimated using the Longuet-Higgins equation (1970a, 1970b),

$$q_1 = 20.7S\sqrt{gh_b} \sin\alpha\cos\alpha, \tag{10.1}$$

where S is the slope of the beach, g is $9.8\,\text{m/s}^2$, h_b is breaker height, and α is the angle the wave makes with the shoreline. At Huntington Beach, CA, q_l typically varies from 0 to $1\,\text{m/s}$. Low littoral drift speeds arise when waves break perpendicular to the shore, high speeds result when waves impinge upon the shoreline at oblique angles. In the simple case of a straight sandy beach bordered offshore by a continental shelf, the littoral surf zone currents take the direction of the alongshore component of the approaching waves. When the momentum of a point source is less than the momentum of the surf zone, the alongshore currents in the surf zone will determine the direction at which pollution from the source will be advected (Inman and Bush, 1973).

Cross-shore exchange between the surf zone and offshore waters is mediated by waves and rip currents that form intermittently along the shoreline. Offshore waters enter the surf zone via breaking waves, and surf zone waters can exit the surf zone via rip currents. When FIB sources are land-based, offshore waters tend to contain lower (if any) concentrations of FIB than waters within the surf zone. Hence, vigorous cross-shore exchange relative to alongshore transport can facilitate a rapid movement of FIB-impacted waters away from the surf zone where bather contact occurs and decrease the length of shoreline adversely impacted by a fecal point source (Boehm, 2003; Grant et al., 2005; Jeong et al., 2005). The conditions under which vigorous cross-shelf exchange occurs are unique to each beach. However, the exchange is greatest when waves impinge upon the shoreline at a perpendicular angle of attack (Longuet-Higgins, 1970a, 1970b; Inman and Bush, 1973).

At freshwater beaches, wind waves (of the order of 10–20 cm) are responsible for resuspending sand/sediment in the swash zone. Once suspended, the sand can limit the penetration of light into the water, and thus prevent the light-induced decay of fecal organisms (see Section 10.2.1). Foreshore sand can be a reservoir of fecal organisms, thus when they strike the swash zone, they may suspend sediment-associated fecal organisms (Alm et al., 2003; Whitman and Nevers, 2003).

10.2.5   Wind

Wind strongly influences the movement of water outside the surf zone of a beach. Thus, winds from a particular direction may be able to transport FIB-laden material to or away from the shoreline. Wind events can also cause waves that may impact FIB densities as outlined above. In the sea, alongshore winds may cause upwelling events during which deep, cold, nutrient-rich waters move to the surface and impinge upon the shoreline. Changes in physical properties of nearshore waters accompanying upwelling events may have an effect on levels of FIB there. For example, cooler temperatures promote prolonged persistence of FIB (Burkhardt *et al.*, 2000; Wait and Sobsey, 2001). Phytoplankton and algal blooms may be initiated by the nutrient-rich upwelled waters. As discussed in Section 10.2.7, exudates or the organisms themselves may serve as substrates upon which FIB may exist extraenterically. Zooplankton blooms may follow phytoplankton blooms, and the exoskeletons of these organisms may also serve as substrates for the extraenteric persistence of FIB (Signoretto *et al.*, 2004).

10.2.6 Temperature

In cooler waters, FIB can persist in a culturable state for longer periods than in warm waters (Noble *et al.*, 2004). Wait and Sobsey (2001) show the same trend for pathogens including *Shigella sonnei*, *Salmonella typhi*, and poliovirus 1. Water temperature may also serve as a marker for water masses with different potentials for microbial pollution or physical processes that impact fecal indicator levels. For example, effluent from an urban drain or stream may be relatively warm, or upwelled waters that may contain elevated levels of nutrients or algae may have cooler temperatures. Internal waves, present in both marine and lacustrine systems and observable as temperature fronts, can induce turbulence at the sediment water interface, potentially mobilizing bacteria present in the sand or sediments.

10.2.7 Biological factors

FIB serve as indicators for human waste, but research illustrates that there are biological reservoirs for these organism that are non-human. The presence of these biological reservoirs in recreational waters or upon recreational beach sands may increase concentrations of FIB in recreational waters.

Plant wrack

Anderson *et al.* (1997) found that ENT concentrations on degrading drift seaweed exceeded those in seawater by 2–4 orders of magnitude at marine recreational beaches in New Zealand. High concentrations of EC and ENT have been reported in *Cladophora glomerata* along the Lake Michigan shore (Whitman *et al.*, 2003). Byappanahalli *et al.* (2003) showed that algal leachate from *Cladophora* readily supported the growth of these indicator organisms, further suggesting that *Cladophora* provides a suitable environment for FIB to persist or grow under natural conditions. Both EC and ENT survived for over six months in sun-dried *Cladophora* mats stored at 4 °C and the residual bacteria in the dried alga readily grew upon rehydration.

Birds

Bird feces contain high levels of FIB. Numerous studies have documented the potential association of FIB in bathing waters with the presence of birds (Oshiro and Fujioka, 1995; Jones and Obiri-Danso, 1999; Grant *et al.*, 2001; Choi *et al.*, 2003; Fogarty *et al.*, 2003). Choi *et al.* (2003) used antibiotic resistance patterns to determine that 8–66 % of the ENT isolates from a polluted marine beach were of bird origin. Thus, bird number may relate to bacterial contamination of recreational waters.

Zooplankton

Signoretto *et al.* (2004) showed that ENT can persist in a non-culturable state associated with zooplankton in seawater and freshwater. Thus, high concentrations of zooplankton or physicochemical conditions that promote their number could be correlated to elevated ENT at beaches.

Human

Bather shedding is another potential source of FIB and pathogens to bathing waters. Papadakis *et al.* (1997) found that concentrations of fecal bacteria and yeasts of human origin in beach sands and water were correlated with the number of visitors present. Sherry (1986) showed that bather load, amongst other parameters, may influence FIB levels in waters of a freshwater beach.

10.2.8   Summary

Based on scientific findings discussed above, it is recommended that the following parameters be recorded during the collection of water samples for FIB analyses in order to serve as inputs to beach water quality models:

   (i)   time of collection;

  (ii)   precipitation;

 (iii)   tide level and range (could be inferred from time of sample collection using tide charts and predictions available from NOAA and NASA among other resources);

 (iv)   wave size and direction;

  (v)   wind speed and direction;

 (vi)   bird numbers;

(vii)   amount of biological wrack;

(viii) number of bathers;

 (ix) temperature.

Items (i)–(viii) can be recorded by a trained individual by observation alone without the use of sophisticated equipment or sensors.

10.3 Models

Various types of models can be applied to beach water quality. *Simple intuitive models* are appropriate if a beach manager anticipates poor water quality (e.g., if there is a nearby release of sewage). This type of model is generally derived from experience or prudence. *Regression models* can be created using factors to predict fecal pollution concentrations if an adequate data set exists. Examples of regression models include analysis of variance, multiple discriminant analysis, multiple regression, tree regression, canonical correlation, principle component analysis, and partial least-squares regression (Hou *et al.*, 2006). Linear regression models assume linear relationships between factors, or combinations of factors, and FIB (Nevers and Whitman, 2005). Classification and regression trees are non-parametric, non-linear regression models. Regression models can have either continuous or binary outputs. *Deterministic models* utilize material and momentum balances to predict the fate and transport of fecal pollutants from well-defined sources (Connolly *et al.*, 1999; Steets and Holden, 2003; Boehm *et al.*, 2005; Grant *et al.*, 2005).

In this chapter, we focus on regression models and their application to beach water quality. Well-developed deterministic models would be extremely useful for modeling beach water quality. However, because our understanding of the complex interplay between physics and microbiology in nearshore environments is limited, deterministic models can only be applied to beaches where the advection, dispersion, sources, and sinks of FIB have been thoroughly characterized by intensive field campaigns. At the present time, regression models are more easily applied to beach water at most beaches.

Once a regression model is constructed it is important to describe its usefulness or success. In this chapter, various metrics are employed for describing the suitability of a regression model for predicting beach water quality. A regression model is built using a 'training' data set comprised of dependent and independent variables. The ability of the model to predict the dependent variable using independent variable inputs within the training data set is described by a root mean square error (RMSE)

$$RMSE = \sqrt{\frac{1}{N} \sum_i \left(\hat{Y}_i - Y_i \right)^2}, \tag{10.2}$$

where $\hat{Y}_i$ and Y_i are the predicted and actual dependent variable, respectively, N is the number of data points used in the training data set, and the index i ranges from 1 to N. A coefficient of determination (R^2) can also be used and is interpreted as the proportion of the variation of the independent data set described by the model. A third metric for testing the performance of a model is to examine the number of Type I and Type II errors that result. Assuming the null hypothesis is that a beach is in compliance with a water quality regulation and should be open, a Type I error occurs when a beach is closed or posted with a warning when it should not be, while a Type II error occurs when a beach is not posted or closed when it should be based on the water quality regulation. These two types of errors can be summed to give the total errors. The number of such errors is a function of the specific policy utilized by beach managers in making posting and closure decisions.

A model must be validated using a data set with which it was not trained before it can be applied as a predictive tool. Validation can only be completed if an appropriate validation data set of independent and dependent variables not used to train the model is available. The success of a model during validation is described by the root mean square error of prediction (RMSEP), which has the same mathematical form as equation (10.2). The number of Type I and II errors, as well as the total error rate is also calculated.

10.4   Regression model case studies

Regression models can potentially improve predictions of FIB concentrations, but they are not yet in widespread use by beach managers. Here we present case studies for two beaches where data are available to construct and evaluate the performance of such models. Three types of continuous regression models (multiple regression, partial least-squares regression, and regression trees) and two types of categorical regression models (discriminant analysis and categorical tree regression) are illustrated using a subset of factors described in the previous section as independent variables. Each of the five regression models is applied to a single beach.

The results from these models are then compared to those from the 'existing model' used by beach managers, in which beach water quality on day T is assumed to be equivalent to water quality on day $T - 1$. This evaluation is accomplished using independent validation data, when available, to assess whether the regression models improve upon water quality predictions made with the existing model.

10.4.1   63rd Street Beach, Chicago: description and data used

63rd Street Beach is a freshwater beach located along Lake Michigan in Chicago (Figure 10.1(a)). The beach is 600 m long with a bathing season from May through September. Concentrations of EC measured in 90 cm of water at 1 p.m. were used as the dependent variable, while wave height, wind speed, sunlight intensity, rainfall, barometric pressure, wave period, and air temperature were available as independent factors. Data were collected from local weather stations, buoy 45007 located in southern Lake Michigan and operated by NOAA, wave measurements from the Army Corps of Engineers, and hydrometeorological information from instruments

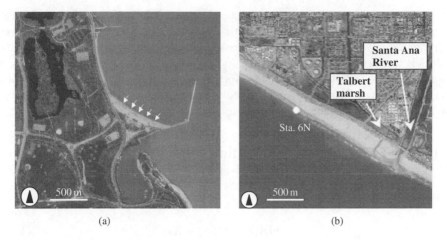

(a)   (b)

Figure 10.1   (a) 63rd Street Beach, Chicago, on the shore of Lake Michigan. Sampling transect sites are shown with white arrows. (b) Huntington State Beach, CA, on the shore of the Pacific Ocean. Maps courtesy of USGS seamless database. Data available from U.S. Geological Survey/EROS, Sioux Falls, SD.

placed at the beach. The data set consisted of 42 measurements made between May 27 and September 30, 2000; days on which data were available for the dependent and all independent variables were included.

10.4.2   Huntington State Beach, Huntington Beach, CA: description and data used

Huntington State Beach (Figure 10.1(b)) is a wave-dominated, open-water beach, typical of marine recreational beaches on the US west coast. Daily tide range is between 1 and 2 m during neap and spring tides, respectively. Waves break on the beach face and nearby sandbars, creating a surf zone where most bather contact occurs. Wave height is typically 0.3 to 1 m but may be larger than 2 m during swells. The state beach is bordered by the Santa Ana River and Talbert Marsh outlets to the south. During the years under study, these outlets discharged freshwater during winter rain events only. During dry weather these outlets serve as saltwater tidal wetlands; flood tides transport seawater into their lower reaches and this water is subsequently flushed during ebb tides.

To construct the models, we used a time series of $\log_{10}$ ENT (log-ENT) at station 6N within Huntington State Beach (Figure 10.1(b)) as the dependent variable. Independent variables included log-transformed same-day and 1- to 3-day lagged volumetric discharge from the Santa Ana River, daily tide range, time of sample collection, tide level during sample collection, direction of littoral drift in the surf zone (upcoast and downcoast), atmospheric pressure, wave height and direction, surf zone water temperature at a pier over 2 km away, deviation of water temperature from 11-year norm, average daily wind speed as recorded at NOAA buoy 46025 near Santa Monica, and log-transformed same-day and 1- to 3-day lagged rainfall over Orange County, CA (averaged from measurements at 19 stations in the watershed). Models were constructed for wet and dry seasons of 2000–2001. The wet season was defined as October 2000 through April 2001, while the dry season was defined as May through September 2001. There were a total of 81 and 88 data points available for constructing the wet and dry season models, respectively.

10.4.3   Evaluation of the existing model at Huntington Beach and 63rd Street Beach

The existing model assumes that

$$\text{FIB}(t) = \text{FIB}(t - X) \qquad\qquad (10.3)$$

where $\text{FIB}(t)$ is the FIB level on day t, and X is the lag between sample collection and policy decisions. In the present analysis, we assume that X is equal to 1 day. The RMSEP for this model was computed for 63rd Street Beach (Table 10.1) and Huntington State Beach (Table 10.2) for EC and ENT. These two indicators were chosen for the freshwater and marine beaches since they best correlate to risk of swimmer illness, respectively (Wade *et al.*, 2003). In each case, Y in equation (10.2) was defined as log-FIB. At 63rd Street Beach, RMSEP = 0.596 for the existing

Table 10.1   Summary of model evaluations for 63rd Street Beach.

Model name	Data origination	Train/ validate	RMSE or RMSEP	Type I error rate (%)	Type II error rate (%)	Total error rate (%)
Existing	—	Validate	0.596	22	14	36
Multiple regression	90 cm, p.m.[a]	Train	0.429	5	0	5
Multiple regression	45 cm, a.m.[b]	Validate	0.575	8	26	34
Regression tree	90 cm, p.m.[a]	Train	0.4828	7	2	9
Discriminant function	90 cm, p.m.[a]	Train	—	7	0	7
Discriminant function	90 cm, p.m.[a]	Validate	—	10	0	10

[a] US Geological Survey.
[b] Chicago Park District

Table 10.2   Summary of model evaluations for Huntington State Beach, Station 6N. The RMSEP for the existing model was measured for the entire data set and is given in the top row.

Model name	Season	Train/ validate	RMSE or RMSEP	Type 1 Error Rate (%)	Type II Error Rate (%)	Total Error Rate (%)
Existing	—	Validate	0.742	6	6	12
PLS	Wet	Train	0.438	0	6	6
PLS	Wet	Validate	0.594	2	7	9
PLS	Dry	Train	0.528	0	8	8
PLS	Dry	Validate	0.871	0	7	7
Tree	Wet	Train	—	1	4	5
Tree	Wet	Validate	—	0	14	14
Tree	Dry	Train	—	2	3	5
Tree	Dry	Validate	—	21	4	25

model based on measurements collected between May and September 2000. At Huntington Beach, the RMSEP for the existing model is 0.742 using dry season data collected between April and October 2001. A plot of the actual versus predicted EC and ENT for 63rd Street Beach and Huntington Beach (station 6N), respectively, is shown in Figure 10.2. This model resulted in 8 (22 %) Type I and 5 (14 %) Type II errors out of 36 testable outcomes at 63rd Street Beach, and 3 (6 %) Type I and 3 (6 %) Type II errors out of 47 testable outcomes at Huntington Beach over the same time periods. The RMSEP and Type I and Type II error rates will be used to ascertain the usefulness of new regression models formulated from environmental factors.

10.4.4   Multiple regression model for 63rd Street Beach

A multiple regression model was constructed for 63rd Street Beach using three factors (wave height, wind speed, and sunlight):

$$y = B_0 + B_1 X_1 + B_2 X_2 + B_3 X_3 + e, \tag{10.4}$$

where B_0 is a constant; y is log-EC, where EC is measured in colony-forming units per 100 ml (CFU/100 ml); B_1, B_2, and B_3 are regression coefficients for the predictors wave height, wind speed, and sunlight, respectively; X_1, X_2, and X_3 are the values of the predictors wave height, wind speed, and sunlight, respectively; and e is the residual error of the model. The three factors were chosen using a

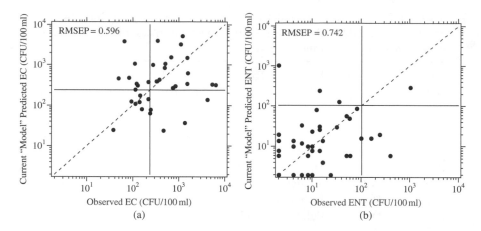

Figure 10.2   (a) Observed versus predicted concentrations of EC at 63rd Street Beach and (b) concentrations of ENT at Huntington Beach, using the existing method of prediction in which EC or ENT is predicted to be equivalent to the previous day's concentration. The dashed line represents the one-to-one line. Note that the existing model cannot be evaluated during the wet season and Huntington State Beach because samples are not collected on consecutive days during the wet season.

stepwise multiple regression that maximized R^2. The analysis yielded the following equation:

$$y = 2.310 + 0.029X_1 - 0.083X_2 - 2.287X_3. \tag{10.5}$$

The positive coefficient for wave height suggests that larger waves give rise to higher EC concentrations. The negative coefficients for wind speed and sunlight indicate that high winds and high sunlight intensity lead to reduced concentrations of EC. The model explains 60 % ($R^2 = 0.597$) of the variation of log-EC in the training data set. This model resulted in an RMSE $= 0.429$ with 2 (5 %) Type I errors and 0 (0 %) Type II errors (Table 10.1). Figure 10.3 shows the predicted versus observed EC.

When constructing a multiple regression model, care must be taken not to violate any of the assumptions used in formulating the model. One such assumption requires the dependent variable not to be autocorrelated, which can be assessed using the Durbin–Watson statistic. In the model above, the Durbin–Watson statistic (based on $\alpha = 0.05$), suggests that log-EC are not autocorrelated. Another assumption requires that the independent variables used to formulate the model are not correlated. This is especially important if the model is to be used for extracting information about relationships between dependent and independent variables. There were no correlations among wind speed, sunlight, and wave height high enough to violate the assumption.

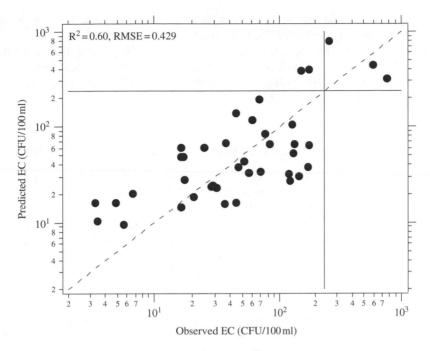

Figure 10.3   Predicted versus observed EC using a multiple regression model. The observed values are the values used to train the model.

For the training data set, the regression model reduced overall error rate from 36% to 5%, and Type II errors from 14% to 0%, compared to the existing model. However, models should be validated with data that were not used in model construction before they are used with confidence. Empirically derived models tend to be biased because they are based on the data that they seek to predict. Cross-validation techniques, such as jackknife procedures, can also be biased since the generated subsets are taken from the data used for model development. Validation is best conducted using independently collected data. To validate the 63rd Street Beach multiple regression model, we used EC data collected by the Chicago Park District (CPD). These data were not ideal for validation, as they were collected at 10 a.m. in 45 cm deep water, whereas our model was developed for EC at 1 p.m. in 90 cm of water. The CPD measurements of EC were generally higher than our measurements on the same days, as would be expected for shallower samples collected earlier in the day, but the two sets of measurements were correlated.

The coefficient of determination for CPD morning samples using the model described in equation (10.5) was only 0.124, and the model was not significant ($p = 0.098$). The RMSEP was 0.575, lower than the existing model, suggesting it is better at predicting EC densities. Using the regression model in place of the existing model would have reduced the number of Type I errors to 8%, but would have increased the number of Type II errors to 26%. The total error rate would have been reduced, but only slightly, to 34% (Table 10.1). Thus, the model did not validate completely. The lack of validation suggests that our model needs some refinement, and also emphasizes that validation data sets should be collected at similar locations and times of day.

10.4.5   Partial least-squares model of Huntington State Beach

We used a partial least-squares regression (PLS) model to examine how physical independent variables can be used to predict ENT levels at station 6N during the wet season of 2000–2001 and the dry season of 2001.

PLS is a combination of multivariable regression, principal components analysis, and canonical correlation analysis. It is a useful regression tool for data sets with a small number of observations with many multicollinear factors. An important feature of PLS algorithms is they do not require that the independent data sets be uncorrelated. We used the Simca-P software (Umetrics, Kinnelon, NJ) for this analysis.

PLS returns principle components extracted from the independent variables, the coefficients that describe the contribution of each independent variable to the principle components, and the 'variable importance on projection' (VIP) values for each independent variable. Whereas in the traditional multi-regression model the magnitude of the regression coefficient is used to evaluate the importance of each variable, in PLS the VIP is used to describe the importance of each predicting variable to the model. Wold (1995) suggests that if the VIP is greater than 0.8, then it should be deemed influential on the dependent variable and retained in a PLS model.

Wet season

PLS extracted two principle components from the data that explain 45% of the variation in log-ENT in the training data set. The PLS-predicted ENT densities are plotted versus the observed ENT densities in Figure 10.4(a). The RMSE for the model is 0.438, and the model gives rise to 0 (0%) and 5 (6%) Type I and Type II errors out of 81 testable outcomes. Note that these evaluations apply to the model and the training data set (Table 10.2).

Table 10.3 summarizes how each of the independent variables contributes to the two principle components. A VIP classification of 1, 2, and 3 was assigned to each variable if VIP$\geq$1, 0.8$\leq$VIP<1, and VIP<0.8, respectively. Thus a VIP classification of 1 indicates the variable is *very influential* in determining ENT densities at 6N, whereas a variable classified as 2 is *moderately influential*. A VIP classification of 3 indicates the factor is *weakly influential*. The sign of the coefficient is given for instances when the coefficient was not within one standard

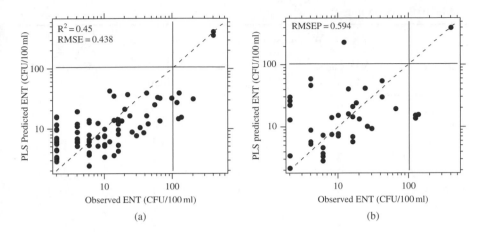

Figure 10.4 (a) Predicted versus observed ENT using the PLS model trained with the entire winter at 6N. The observed values are the values used to train the model. (b) Predicted versus observed ENT validation. The PLS model was trained with a group of wet season data not shown here, and then applied to wet season data that the model had not been trained with.

Table 10.3 VIP classifications and sign of the coefficient in the wet and dry season models for Huntington Beach Station 6N. A VIP classification of 1 indicates VIP ≥ 1 (very influential), 2 indicates $0.8 \leq$ VIP<1 (moderately influential), 3 indicates VIP<0.8 (weakly influential). A blank in the VIP column indicates that the variable was not used in the model for that season. A blank in the coefficient column indicates that the coefficient value was not significantly different from 0. Temperatures are water temperatures. SAR is the Santa Ana River, T is water temperature, S, SSW, SW, W, WSW are south, south-southwest, west, and west-southwest, respectively.

Variable	Wet season		Dry Season	
	VIP classi- fication	Coefficient sign	VIP classi- fication	Coefficient sign
SAR	1	(+)	3	(−)
SAR 1-day lag	1	(+)	—	
SAR 2-day lag	1	(+)	—	
SAR 3-day lag	1		—	
Littoral drift	1	(+)	2	
Rain	1	(+)	1	(−)
Rain 1-day lag	1	(+)	—	
Rain 2-day lag	1	(+)	—	
Rain 3-day lag	3	(+)	—	
Rain 4-day lag	3	(−)	—	

Table 10.3   (Continued)

Variable	Wet season		Dry Season	
	VIP classi-fication	Coefficient sign	VIP classi-fication	Coefficient sign
Pressure	2	(−)	3	
Time	1	(+)	1	(−)
Daily T	2	(+)	1	(+)
Daily T dev.	2	(+)	1	
Morning T	2	(+)	1	(+)
Tide range	3	(+)	1	(+)
Tide level	2	(+)	3	
S wave	1		2	(−)
SSW wave	3		3	(+)
SW wave	3		1	(+)
W wave	3	(−)	3	
WSW wave	3		3	
Wave height	3		3	

error of zero. The standard error of the coefficient was determined by a jack-knife validation (leave-*n*-out method) by Simca-P.

Based on these results, discharge from the Santa Ana River and its lagged discharge were very influential and positively related to log-ENT at station 6N. The exception was the 3-day lagged discharge whose coefficient was not significantly different from 0. Littoral drift was also very influential, with log-ENT responding positively to upcoast littoral drift. Rain, and the 1-day and 2-day lag rain, were very influential, producing a positive response in log-ENT. Finally, time of sample collection was very influential, with later collection times producing higher log-ENT levels. This could potentially have been an artifact of the sampling procedure used by the monitoring agency during storm events. When there was a storm, sampling was typically done later in the day. The presence of waves from the south ranked as very influential according to the magnitude of its VIP; however, its coefficient was not significantly different from zero. Without considering the standard errors, the model predicts a positive coefficient for waves, which would be consistent with our understanding of littoral drift in the system: waves from the south produce upcoast currents that facilitate the discharge from the watershed outlets to the south and to the beach we are modeling. The remaining variables were considered to be moderately and weakly influential and will not be discussed further.

The PLS model described above was retrained with only half of the wet season data set (1 October 2000 through 23 January 2001). The new model shares similar properties as the wet season model constructed with all the winter observations. The model was subsequently validated using the second half of the wet season (24 January through 30 April 2001). Figure 10.4(b) shows the observed versus predicted ENT of the validation. The RMSEP for the validation is 0.594, which is less than the existing wet season benchmark of 0.742. The model gives rise to 1 (2 %) Type I and 3 (7 %) Type II errors, respectively, out of 41 testable outcomes (Table 10.2). Use of the PLS model would reduce the total error rate (9 % compared to 12 % for the existing model). However, while the Type I error rate is decreased, the Type II error rate is slightly higher relative to the benchmark. It should be noted that the existing model that we seek to improve has been validated during the dry season only. Because samples are not collected on consecutive days at Huntington State Beach during the wet season, it is impossible to evaluate the actual success of the 'existing model' during the wet season.

Dry season

The dry season actually included a few rainfall events; thus, rainfall and Santa Ana River discharge were used as factors in the model. However, their lagged time series were not included. Using the data from the entire dry season, the PLS model produced two principle components that together explained 34 % of the variation in log-ENT levels of the training data set. The PLS-predicted ENT densities are plotted versus the observed ENT densities in Figure 10.5(a). The RMSE for the model is 0.528, and the model gives rise to 0 (0 %) Type I and 7 (8 %) and Type II errors out of 88 testable outcomes in the training data set (Table 10.2).

Table 10.3 shows the VIP classification and sign of the corresponding coefficients when they are significantly different zero (using the same criteria described above). The very influential factors include rain, time of sampling, daily average water temperature, morning water temperature, water temperature deviation from 11-year norm, tide range, presence of waves from the south, and presence of waves from the southwest. Of these, the coefficient for temperature deviation from 11-year norm is not significantly different from zero, indicating that its contribution to the principle components, and therefore log-ENT, is uncertain. The role of rain and Santa Ana River discharge in the dry season model is different than during the wet season because there is very little rain and, consequently, discharges from the river. There were three small rain events in the beginning of the dry season. The negative coefficient for rain can be explained by the fact that log-ENT is actually higher in the dry season when there is no rain, suggesting rain-induced runoff is not a principle source of ENT at the beach during the dry season. The negative coefficient for time of sampling is consistent with our understanding of sunlight-mediated inactivation of ENT during dry weather (sunlight causes die-off of ENT). The other very influential factors have positive coefficients, indicating that log-ENT responds positively to the presence of waves from the SW, spring tides, and warmer water temperatures. Waves from the SW are very common during the summer at the field site (Boehm, 2003), and these produce upcoast littoral currents, which potentially transport tidal discharge from the river mouth to the beach. Spring tides bring large tide ranges that have been shown to coincide with episodes of pollution at southern California beaches (Boehm and Weisberg, 2005). The relationship between ENT and water temperature is consistent with the results for rainfall. Most of the pollution events are in mid-summer when there is no rainfall and the waters are the warmest of the season. It should be noted that the water temperature used in this model has not been high pass filtered, so it contains a strong seasonal signal. Thus, the relationship between water temperature and water quality in this model cannot be used to make conclusions about the relationship between water quality and synoptic or tidal cooling discussed in detail by Boehm *et al.* (2004). The remaining factors are moderately and weakly influential based on their VIP values and will not be discussed here.

To validate the dry season model, we constructed a PLS model using data from half of the dry season (1 May through 18 July 2001) and used the next two months (19 July through 30 September 2001) to validate the model. Figure 10.5(b) shows the observed versus predicted data. The RMSEP (given in the upper corner, 0.871) is larger than that of the existing model (RMSEP = 0.742), suggesting that this PLS model is not superior to the existing model for predicting daily ENT densities. This implies that we have not included all the factors that influence ENT in the model or that the way the factors influence ENT varies from month to month. The validation predicts 0 (0 %) Type I and 3 (7 %) Type II errors out of 41 outcomes. This is a lower overall error rate (7 %) than the current dry weather model (12 %).

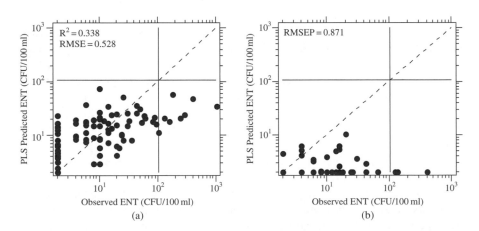

Figure 10.5   (a) Predicted versus observed ENT using the PLS model trained with the entire dry season at 6N. The observed values are the same values used to train the model. (b) Predicted versus observed ENT validation. The PLS model was trained with a group of dry season data not shown here, and then applied to dry season data that the model had not been trained with.

Type I errors are lower (0 % compared to 6 %), but Type II errors are higher (7 % compared to 6 %) (Table 10.2).

10.4.6   Regression trees

Regression trees represent a non-parametric, non-linear method of relating independent variables to dependent variables (Breiman *et al.*, 1984). Interaction terms between independent variables are detected automatically during tree construction. Regression trees have been applied to modeling ecological data (Rejwan *et al.*, 1999; De'ath and Fabricius, 2000) and beach water quality (Parkhurst *et al.*, 2005). They are relatively simple to interpret because they schematically represent nested *if – then* relationships between variables. They are formed by splitting dependent variable groups using independent variable criteria so that the data within groups are more similar than data split between groups. In theory, the same number of groups as there are data points can be created so that the regression tree is perfectly capable of 'predicting' the training data set. It is preferable to implement a stopping rule so that trees do not have too many branches and leaves.

A regression tree algorithm (SYSTAT, Point Richmond, CA) was used to create a regression tree for 63rd Street Beach (Figure 10.6) using the three independent variables used in the multiple regression modeling. The data were sorted into homogeneous subsets using recursive partitioning. A stopping rule of $N = 5$ in terminal nodes was imposed. At the end of each branch are leaf nodes, which here are represented by compartments containing the mean, standard deviation, and total number of cases sorted. Above each node is the binary criteria used to split the data.

The top box represents the total number in the sampled population. The initial population of 42 observations had a $\log_{10}$-mean (log-mean) EC of 1.749 CFU/100 ml (note the units CFU/100 ml following a log-mean refer to the units of EC concentration). The first branching is based on wave height. When wave height was below 31 cm, the log-mean was 1.52 CFU/100 ml, and above this criterion, the log-mean was 2.33 CFU/100 ml (thus, many of the readings in this subset will be in exceedance of the EC water quality standard log-EC equivalent to 2.37 CFU/100 ml). The next branching uses a wind speed criterion of 14.4 m/s. When wind speed is below this value and wave height above 31 cm, the log-mean was 2.577 CFU/100 ml. Apparently higher offshore winds had a moderating effect on EC concentrations while slower winds (during increased waves) allowed contaminants to accumulate in the nearshore water. At low wave heights, sunlight was important, but both 'leaves' of the tree are well below beach closing criteria. This is apparently because of the negative effect of sunlight on the bacteria coupled with enhanced exposure by decreased turbulence, turbidity, and surface conditions. Overall, the proportion reduction in error (equivalent to the R^2) for this tree was 47 % and approaches that delivered by traditional linear regression models, with comparable validation using jackknife analysis. For the tree, the RMSE was 0.4828, and the error rate included 3 (7 %) Type I errors and 1 (2 %) Type II error, for a total 9 % error rate. By way of comparison, the currently used model had an RMSE of 0.596 with a total error rate of 36 % (Table 10.1). Note that the success of this tree has been evaluated using the model and the training data set.

10.4.7   Binary outcome models

Often a modeler is less interested in accurately predicting FIB concentration than in accurately predicting when the beach is in or out of compliance with water quality criteria. Models specifically formulated for categorical dependent variables should be used when one wishes to model a binary categorical variable describing beach water quality (in or out of compliance). Possibilities include logistic regression, discriminant analysis, and categorical regression trees (Tabachnick and Fidell, 1996). In these models, independent variables can be continuous or categorical.

Whether a beach is open or closed/posted depends on the specific policy in place. At 63rd Street Beach, it is assumed that a beach with EC<235 CFU/100 ml should be open, whereas if EC≥235 CFU/100 ml it should be closed. At Huntington Beach, warning signs are posted when the ENT standard 104 CFU/100 ml is exceeded, with closure reserved only for circumstances when there is a known sewage spill. In the following sections, we will illustrate two methods to predict binary outcomes: discriminant analysis and categorical tree regression.

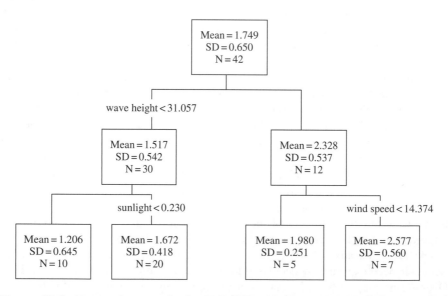

Figure 10.6   Regression tree for log-EC (EC with units of CFU/100 ml) at 63rd Street Beach. The mean describes the average log-EC in each box. Wave height is reported in centimeters, sunlight in langleys per second, and wind speed in meters per second.

10.4.8 Binary outcome models: discriminant analyses

For 63rd Street Beach, a multiple analysis of variance was performed to determine the discriminant functions for the same data set used for the multiple regression. We first tested the significance of the derived discriminant functions and then classified the EC data as in compliance or out of compliance based on those functions. The following set of discriminant function coefficients was highly significant ($p < 0.0001$, Wilks' lambda $= 0.497$, df $= 3$): wind speed, sunlight, and wave height.

Using the discriminant function, we were able to predict correctly 34 out of 37 (92%) of the EC observations that were within compliance with the single-sample standard, and 5 out of 5 (100%) of the out-of-compliance observations. This model gave rise to 3 (7%) Type I and 0 (0%) Type II errors out of 42 testable outcomes (Table 10.1). Note that these results were obtained by comparing the model with the observations with which it was trained.

We validated the categorical model by rebuilding it with a smaller training data set and validating with the remaining observations within the original training data set. Thirty-three of 37 (89%) and 5 of 5 (100%) of the in-compliance and out-of-compliance observations were correctly predicted, respectively. Four (10%) and 0 (0%) out of 42 testable outcomes were Type I and Type II errors, respectively. Discriminant analysis successfully eliminated many of the Type I and Type II errors made by the existing model (recall, Type I and II errors occurred at rates of 22% and 14% in the existing model, respectively).

10.4.9 Binary outcome models: categorical tree analysis

We used a categorical tree to model whether Huntington Beach at station 6N is in compliance with the ENT single-sample water quality standard (ENT <104 CFU/100 ml, assigned a value of 0) or out of compliance with the standard (ENT ≥ 104 CFU/100 ml, assigned a value of 1). We constructed a regression tree for wet and dry seasons separately using only a subset of the factors used for PLS.

To create the trees, we utilized the categorical version of the regression tree in SYSTAT. The technique is similar as that used for the regression tree described earlier for 63rd Street Beach, but the compartments contain a categorical variable, the number of cases, and an impurity rather than a log-mean FIB concentration, the number of cases, and standard deviation. The impurity is defined as the product of the probability of category 0 at node i and the probability of category 1 at node i,

$$\text{impurity} = \frac{n_1(N - n_1)}{N^2}, \tag{10.6}$$

where N is the number of observations at a node and n_1 is the number of observations of the categorical variable 1. If impurity $= 0$, then all the observations portioned into the compartment are completely homogeneous. If impurity $= 0.25$, then half the observations at the node are 1 and half are 0, implying the node is completely heterogeneous.

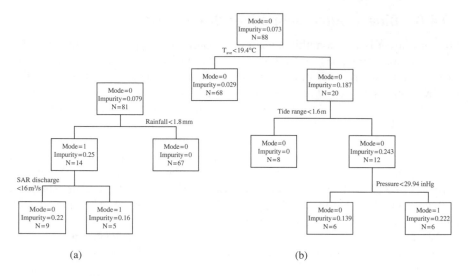

(a)   (b)

Figure 10.7   (a) Wet and (b) dry season categorical trees for station 6N at Huntington State Beach.

Figure 10.7 shows categorical regression trees for station 6N during wet and dry seasons. The box at the top of each tree shows the total number in the sampled population.

Wet season

During the wet season, there are 81 observations to partition (Figure 10.7(a)). The first criterion used to partition the data was rainfall at a threshold of 1.8 mm. The 67 data points collected when rainfall was less than 1.8 mm were all in compliance (impurity $= 0$). The remaining 14 days were further partitioned based on a threshold discharge from the Santa Ana River of 16 m³/s. The majority (6 of 9) of the samples collected when the discharge was less than 16 m³/s were in compliance with the single-sample standard (mode $= 0$, impurity $= 0.22$). The majority of the remaining samples (4 of 5) collected when discharge was greater than 16 m³/s were out of compliance with the standard (mode=1, impurity=0.16). Overall, an exceedance of the single-sample standard is expected to occur when rainfall is greater than 1.8 mm and river discharge exceeds 16 m³/s. The importance of rainfall and discharge is also borne out by the PLS wet season model. Overall, this model correctly categorized 73 of 74 (99 %) and 4 of 7 (57 %) observations of ENT in and out of compliance with the single-sample standard, respectively. One (1 %) Type I and 3 (4 %) Type II errors, respectively, were predicted out of 81 testable outcomes (Table 10.2). Note that this description is for the unvalidated model.

To validate the model, the same approach was applied to data collected during the 2001–2002 wet season (November 2001 through April 2002). There are 105 observations available for the validation. During this period, there were 15 single-sample standard exceedances observed. Using the decision tree in Figure 10.7(a), no exceedances were predicted. This implies that there were 0 (0%) Type I and 15 (14%) Type II errors, respectively. Although the percentage of Type I errors is less than our benchmark (the current lag model), the percentages of Type II errors and total errors are greater (Table 10.2). This implies that the categorical tree could be improved by including more variables, or needs to be built using observations from the same season as those used to validate the model.

Dry season

During the dry season, the average water temperature, tide range, and atmospheric pressure were used as criteria for creating the classification tree of 88 samples (Figure 10.7(b)). The first branching threshold is a daily average water temperature of 19.4 °C. When the water is cooler than the threshold temperature, 68 samples are partitioned into a compartment where the majority (66 of 68) are in compliance with the standard (mode=0, impurity=0.029). The remaining samples are further partitioned based on tide range. When tide range is less than 1.6 m (neap tides), 6 samples are compartmentalized that are within compliance with the standard (mode=0, impurity=0). The remaining samples collected when the tide range is greater than 1.6 m (spring tides) are further partitioned using an atmospheric pressure threshold of 29.94 inHg. When the pressure is less than this threshold (implying cloudy conditions), 6 samples are partitioned into a compartment where the majority (4 of 6) are out of compliance with the standard (mode=1, impurity=0.22). Interpreting all the branches together, exceedance of the ENT single-sample standard is most likely to occur when water temperature is greater than 19.4 °C, tide range is greater than 1.6 m, and atmospheric pressure is lower than 29.94 inHg. The description of conditions for partial to poor water quality agrees qualitatively with the interpretation of the dry weather PLS model described above.

Overall, the model correctly categorized 79 of 81 (98%) and 4 of 7 (57%) observations of ENT in and out of compliance with the single-sample standard, respectively. Two (2%) Type I and 3 (3%) Type II errors, respectively, were predicted out of 88 testable outcomes (Table 10.2). Note that this description is for the unvalidated model.

To validate the model, we applied it to data collected during the dry season of 2000. There are 72 observations available for the validation. During this period, there were 9 exceedances observed in the single-sample standard. Based on the decision tree in Figure 10.7(b), 18 exceedances were predicted, 3 of which were correct. This implies that there were 15 (21%) Type I and 3 (4%) Type II errors, respectively (Table 10.2). Although the Type II errors are less than our benchmark (the current lag model), the Type I errors and the total number of errors are greater. This implies that the categorical tree could be improved by including more variables, or needs to be built using observations from the same season as those used to validate the model.

10.5   Summary of modeling results

Tables 10.1 and 10.2 summarize the criteria used to evaluate the current and regression models at 63rd Street Beach and Huntington State, respectively.

At Huntington State Beach, the wet season PLS models performed better than the existing model based on a comparison of the RMSEP and total error rate of the PLS wet season validation and the existing model. While the Type I error rate decreased (clean beaches were closed incorrectly less often) using the PLS wet model, the Type II errors were slightly more common (beaches were predicted to be in compliance when they were not). The dry weather PLS model better predicted day-to-day ENT levels within the training data set compared to the existing model based on the RMSE. Total error rates were also lower, as were Type I errors. The dry season model did not do better than the existing model under validation at predicting day-to-day variations in ENT. However, the total error rate was lower than the existing model; Type I errors were non-existent, while Type II errors increased slightly.

The regression trees did better at predicting the training data set ENT values than they did with validation data sets. The wet season tree was more successful than the dry season based on a comparison of error rates to the existing model.

The results from the PLS and the regression tree models at Huntington State Beach suggest that we need a better understanding of the factors or random variables that influence fate and transport of ENT, especially during the dry season at this marine beach, in order to construct superior models. As they stand now, these models perform better, under at least one of the evaluation criteria, than the existing model used by beach managers.

At 63rd Street Beach, the regression model performed better than the existing model with the training data set. Type II errors were eliminated after occurring 22 % of the time using the existing modeling approach. The model validations performed poorly, but this was probably due to the lack of an appropriate validation data set. The validation data were collected earlier in the day and in shallower water, both of which lead to higher EC counts than the data used to develop the model. These results underscore the importance of collecting validation data sets during modeling studies.

10.6   Conclusions

We illustrate applications of both continuous and categorical regression models to beach water quality data at freshwater and marine beaches. While it was possible to train models to predict FIB densities at beaches, the models did not consistently validate well. Based on the validation criteria (RMSEP and Type I and II errors), some of the models presented in this chapter do not improve drastically on the existing model. To improve models, we need to learn more about FIB sources, natural variation, and the processes that govern FIB fate and transport in the very near shore environment. This can only be accomplished through interdisciplinary studies that characterize both the extraenteric ecology of FIB and circulation of the very nearshore coastal zone.

In some cases, the models presented here could not be adequately validated due to lack of an additional validation data set. The collection of validation data is just as important as the collection of training data. Predictive models of beach water quality cannot be confidently applied as a policy tool without this additional confirmation and calibration information.

10.7   The future of modeling

While these models show promise for improving on the temporal extrapolations that are presently used in beach warning systems, work is still necessary to improve them further. Models are likely to improve in the future as data on environmental factors and FIB concentrations become more readily available. There are a number of initiatives to develop real-time information about ocean conditions, largely coordinated under the umbrella of ocean observing systems. The implementation of these ocean observing systems has been endorsed by the US Commission on Ocean Policy and is included in the President's Ocean Action Plan.

Ocean observing systems will enhance the number and types of physical oceanographic measurements that are readily available to beach managers. Present plans call for a nationwide system of high-frequency radar deployments that will provide continuous measurements of surface current speed and direction. This type of directional information will particularly enhance predictions at locations where there are known sources of potential fecal contamination, such as stormwater outlets or wastewater outfalls. Observing systems will also involve mooring deployments that will provide real-time data on wave height, water temperature, wind speed, and other nearshore meteorological data. These factors were shown earlier in the chapter to be predictive, even when the data were only available on coarser spatial scales. As the mooring systems become fully operational, these types of data will even become available in a predictive mode, which in turn can be imported into bacterial models to predict bacterial conditions into the future. Such capabilities will allow beachgoers who travel to a site the opportunity to plan their trip based on likely water quality conditions.

Also likely to improve modeling capabilities are advancements in molecular microbiological measurement methods that will soon provide bacterial indicator concentrations in less than 4 hours (Noble and Weisberg, 2005). The models presented in this chapter have focused on using physical data, such as tides, wind, and rain, because those data are available on the day for which a prediction is desired. However, these factors are primarily modifiers of a bacterial concentration based on past conditions. These models would not predict upsets in the source term, such as the occurrence of a sewage spill. The inclusion of more recent bacteria data will protect against the upset scenario and, when combined with meteorological nowcasts and forecasts, will provide a robust predictive capability for beach water quality warning systems.

Acknowledgments

This article is Contribution 1382 of the USGS Great Lakes Science Center. The authors acknowledge Larry Wymer, Sharyl Rabinovici, and Stanley Grant for their thoughtful comments that improved the chapter.

References

Ackerman, D., and Schiff, K. (2003) Modeling stormwater mass emissions to the Southern California Bight. *Journal of Environmental Engineers*, 129: 308–317.

Ackerman, D., and Weisberg, S.B. (2003) Relationship between rainfall and beach bacterial concentration on Santa Monica Bay beaches. *Journal of Water and Health*, 1: 85–89.

Alm, E.W., Burke, J., and Spain, A. (2003) Fecal indicator bacteria are abundant in wet sand at freshwater beaches. *Water Research*, 37: 3978–3982.

Anderson, S.A., Turner, S.J., and Lewis, G.D. (1997) Enterococci in the New Zealand environment: Implications for water quality monitoring. *Water Science and Technology*, 35: 325–331.

Bellair, J.T., Parrsmith, G.A., and Wallis, I.G. (1977) Significance of diurnal variations in fecal coliform die-off rates in design of ocean outfalls. *Journal of Water Pollution Control Federation*, 49: 2022–2030.

Boehm, A.B. (2003) A model of microbial transport and inactivation in the surf zone and application to field measurements of total coliform in Northern Orange County, California. *Environmental Science and Technology*, 37: 5511–5517.

Boehm, A., and Weisberg, S.B. (2005) Tidal forcing of enterococci at marine recreational beaches at fortnightly and semidiurnal frequencies. *Environmental Science and Technology*, 39: 5575–5583.

Boehm, A.B., Grant, S.B., Kim, J.H., Mowbray, S.L., McGee, C.D., Clark, C.D., Foley, D.M., and Wellman, D.E. (2002a) Decadal and shorter period variability and surf zone water quality at Huntington Beach, California. *Environmental Science and Technology*, 36: 3885–3892.

Boehm, A.B. Sanders, B.F., Winant, C.D. (2002b) Cross-shelf transport at Huntington Beach. Implications for the fate of sewage discharged through an offshore ocean outfall *Environmental Science and Technology*, 36: 1899–1906.

Boehm, A.B., Lluch-Cota, D.B., Davis, K.A., Winant, C.D., and Monismith, S.G. (2004) Covariation of coastal water temperature and microbial pollution at interannual to tidal periods. *Geophysical Research Letters*, 31, L06309, doi: 10.1029/2003GL019122.

Boehm, A.B., Keymer, D.P., and Shellenbarger, G.G. (2005) An analytical model of enterococci inactivation, grazing, and transport in the surf zone of a marine beach. *Water Research*, 39: 3565–3578.

Breiman, L., Friedman, J.H., Olshen, R.A., and Stone, C.J. (1984) *Classification and Regression Trees.* Wadsworth, Belmont, CA.

Burkhardt III, W., Calci, K.R., Watkins, W.D., Rippey, S.R., and Chirtel, S.J. (2000) Inactivation of indicator microorganisms in estuarine waters. *Water Research*, 34: 2207–2214.

Byappanahalli M.N., Shively, D.A., Nevers, M.B., Sadowsky, M.J., and Whitman, R.L. (2003) Growth and survival of *Escherichia coli* and enterococci populations in the macroalga Cladophora (Chlorophyta). *FEMS Microbiology Ecology* 46: 203–211.

Choi S., Chu, W.P., Brown, J., Becker, S.J., Harwood, V.J., and Jiang, S.C. (2003) Application of enterococci antibiotic resistance patterns for contamination source identification at Huntington Beach, California. *Marine Pollution Bulletin*, 46: 748–755.

Coelho, M.P.P., Marques, M.E., and Roseiro, J.C. (1999) Dynamics of microbiological contamination at a marine recreational site. *Marine Pollution Bulletin*, 38(12): 1242–1246.

Connolly, J.P., Blumberg, A.F., and Quadrini, J.D. (1999) Modeling fate of pathogenic organisms in coastal waters of Oahu, Hawaii. *Journal of Environmental Engineering – ASCE*, 125: 398–406.

Crowther, J., Kay, D., and Wyer, M.D. (2001) Relationships between microbial water quality and environmental conditions in coastal recreational waters: The Fylde Coast, UK. *Water Research*, 35: 4029–4038.

Curriero, F.C., Patz, J.A., Rose, J.R., and Lele, S. (2001) The association between extreme precipitation and waterborne disease outbreaks in the United States, 1948–1994. *American Journal of Public Health*, 91: 1194–1199.

Davies-Colley, R.J., Bell, R.G., and Donnison, A.M. (1994) Sunlight inactivation of enterococci and fecal coliform in sewage effluent diluted in seawater. *Applied and Environmental Microbiology*, 60: 2049–2058.

De'ath, G., and Fabricius, K.E. (2000) Classification and regression trees: A powerful yet simple technique for ecological data analysis. *Ecology*, 81: 3178–3192.

Fogarty, L.R., Haack, S.K., Wolcott, M.J., and Whitman, R.L. (2003) Abundance and characteristics of the recreational water quality indicator bacteria *Escherichia coli* and enterococci in gull faeces. *Journal of Applied Microbiology*, 94: 865–878.

Fujioka, R.S., and Narikawa, O.T. (1982) Effect of sunlight on enumeration of indicator bacteria under field conditions. *Applied and Environmental Microbiology*, 44: 395–401.

Fujioka, R.S., Hasimoto, H.H., Siwak, E.B., and Young, R.H.F. (1981) Effect of sunlight on survival of indicator bacteria in seawater. *Applied and Environmental Microbiology*, 41: 690–696.

Grant, S.B., Sanders, B.F., Boehm, A.B., Redman, J.A., Kim, J.H., Mrse, R.D., Chu, A.K., Gouldin, M., McGee, C.D., Gardiner, N.A., Jones, B.H., Svejkovsky J., and Leipzig, G.V. (2001) Generation of enterococci bacteria in a coastal saltwater marsh and its impact on surf zone water quality. *Environmental Science and Technology*, 35: 2407–2416.

Grant, S.B., Kim, J.H., Jones, B.H., Jenkins, S.A., Wasyl, J., and Cudaback, C. (2005) Surf zone entrainment alongshore transport, and human health implications of pollution from tidal outlets. *Journal of Geophysical Research – Oceans* 110, C10025.

Haugland, R.A., Siefring, S.C., Wymer, L.J., Brenner, K.P., and Dufour, A.P. (2005) Comparison of *Enterococcus* measurements in freshwater at two recreational beaches by quantitative polymerase chain reaction and membrane filter culture analysis. *Water Research*, 39: 559–568.

Hou, D., Rabinovici, S., and Boehm, A.B. (2006) Enterococci predictions from a partial least squares regression model can improve the efficacy of beach management advisories. *Environmental Science and Technology*, 40: 1737–1743.

Hughes, K.A. (2003) Influence of seasonal environmental variables on the distribution of presumptive fecal coliforms around an Antarctic research station. *Applied and Environmental Microbiology*, 69: 4884–4891.

Inman, D.L., and Bush, B.M. (1973) The coastal challenge. *Science*, 181: 20–32.

Jeong, Y., Grant, S.B., Ritter, S., Pednekar, A., Candelaria, L., and Winant, C.D. (2005) Identifying pollutant sources in tidally mixed systems: Case study of fecal indicator bacteria from marinas in Newport Bay, southern California. *Environmental Science and Technology*, 39: 9083–9093.

Jones, K., and Obiri-Danso, K. (1999) Non-compliance of beaches with the EU directives of bathing water quality: Evidence of non-point sources of pollution in Morecambe Bay. *Journal of Applied Microbiology*, 85: 101S–107S.

Kim, J.H., and Grant, S.B. (2004) Public mis-notification of coastal water quality: A probabilistic evaluation of posting errors at Huntington Beach, California. *Environmental Science and Technology*, 38: 2497–2504.

Kim, J.H., Grant, S.B., McGee, C.D., Sanders, B.F., and Largier, J.L. (2004) Locating sources of surf zone pollution: A mass budget analysis of fecal indicator bacteria at Huntington Beach, CA. *Environmental Science and Technology*, 38: 2626–2636.

Kistemann, T.C., Koch, C., Dangendorf, F., Fischeder, R., Gebel, J., Vacata, V., and Exner, M. (2002) Microbial load of drinking water reservoir tributaries during extreme rainfall and runoff. *Applied and Environmental Microbiology*, 68: 2188–2197.

Leecaster M.K., and Weisberg, S.B. (2001) Effect of sampling frequency on shoreline microbiology assessments. *Marine Pollution Bulletin*, 42: 1150–1154.

Lipp, E.K., Schmidt, N., Luther, M.E., and Rose, J.B. (2001) Determining the effects of El Niño-Southern Oscillation events on coastal water quality. *Estuaries*, 24: 491–497.

Longuet-Higgins, M.S. (1970a) Longshore currents generated by obliquely incident sea waves 1. *Journal of Geophysical Research*, 75: 6778–6789.

Longuet-Higgins, M.S. (1970b) Longshore currents generated by obliquely incident sea waves 2. *Journal of Geophysical Research*, 75: 6790–6801.

McCambridge, J., and McMeekin, T.A. (1981) Effect of solar radiation and predacious microorganisms on survival of fecal and other bacteria. *Applied and Environmental Microbiology*, 41: 1083–1087.

National Research Council (2004) *Indicators for Waterborne Pathogens*. National Academies Press, Washington, DC.

Nevers, M.B., and Whitman, R.L. (2005) Nowcast modeling of *Escherichia coli* concentrations at multiple urban beaches of southern Lake Michigan. *Water Research*, 39: 5250–5650.

Noble, R.T., and Weisberg, S.B. (2005) A review of technologies being developed for rapid detection of bacteria in recreational waters. *Journal of Water and Health*, 3: 381–392.

Noble, R.T., Lee, I.M., and Schiff, K.C. (2004) Inactivation of indicator micro-organisms from various sources of fecal contamination in seawater and freshwater. *Journal of Applied Microbiology*, 96: 464–472.

Obiri-Danso, K., and Jones, K. (2000) Intertidal sediments as reservoirs for hippurate negative campylobacters, salmonellae and faecal indicators in three EU recognized bather waters in north west England. *Water Research*, 34: 519–527.

Oguma, K., Katayama, H., Mitani, H., Morita, S., Hirata, T., and Ohgaki, S. (2001) Determination of pyrimidine dimers in *Escherichia coli* and *Cryptosporidium parvum* during UV light inactivation, photoreactivation, and dark repair. *Applied and Environmental Microbiology*, 67: 4630–4637.

Olyphant, G.A., and Whitman, R.L. (2004) Elements of a predictive model for determining beach closures in a real-time basis: The case of 63rd Street Beach Chicago. *Environmental Monitoring and Assessment*, 98: 175–190.

Oshiro, R., and Fujioka, R. (1995) Sand, soil, and pigeon droppings: Sources of indicator bacteria in the waters of Hanauma Bay, Oahu, Hawaii. *Water Science and Technology*, 33: 251–254.

Papadakis, J.A., Mavridou, A., Richardson, S.C., Lampiri, M., and Marcelou, U. (1997) Bather-related microbial and yeast populations in sand and seawater. *Water Research*, 31: 799–804.

Paranhos, R. Pereira, A.P., and Mayr, L.M. (1998) Diel variability of water quality in a tropical polluted bay. *Environmental Monitoring and Assessment*, 50: 131–141.

Parkhurst, D.F., Brenner, K.P., Dufour, A.P., and Wymer, L.J. (2005) Indicator bacteria at five swimming beaches – analysis using random forests. *Water Research*, 39: 1354–1360.

Rejwan, C., Collins, N.C., Brunner, L.J., Shuter, B.J., Ridgway, M.S. (1999) Tree regression analysis on the nesting habitat of smallmouth bass. *Ecology*, 80: 341–348.

Schiff, K.C., Weisberg, S.B., and Raco-Rands, V.E. (2002) Inventory of ocean monitoring in the Southern California Bight. *Environmental Management*, 29: 871–876.

Schiff, K.C., Morton, J., and Weisberg, S.B. (2003) Retrospective evaluation of shoreline water quality along Santa Monica Bay beaches. *Marine Environmental Research*, 56: 245–254.

Sherry, J.P. (1986) Temporal distribution of faecal pollution indicators and opportunistic pathogens at a Lake Ontario bathing beach. *Journal of Great Lakes Research*, 12: 154–160.

Signoretto, C., Burlacchini, G., del Mar Lleo, M., Pruzzo, C., Zampini, M., Pane, L., Franzini, G., and Canepari, P. (2004) Adhesion of *Enterococcus faecalis* in the nonculturable state to plankton is the main mechanism responsible for persistence of this bacterium in both lake and seawater. *Applied and Environmental Microbiology*, 70: 6892–6896.

Sinton, L.W., Finlay, R.K., Lynch, P.A. (1999) Sunlight inactivation of fecal bacteriophages and bacteria in sewage-polluted seawater. *Applied and Environmental Microbiology*, 65: 3605–3613.

Sinton, L.W., Hall, C.H., Lynch, P.A., and Davies-Colley, R.J. (2002) Sunlight inactivation of fecal indicator bacteria and bacteriophages from waste stabilization pond effluent in fresh and saline waters. *Applied and Environmental Microbiology*, 68: 1122–1131.

Steets, B.M., and Holden, P.A. (2003) A mechanistic model of runoff-associated fecal coliform fate and transport through a coastal lagoon. *Water Research*, 37: 589–608.

Tabachnick, B.G., and Fidell, L.S. (1996) *Using Multivariate Statistics*. HarperCollins, New York.

Taggart, M.L. (2002) Oceanographic and discharge factors affecting shoreline fecal bacteria densities around storm drains and freshwater outlets at marine beaches. DEnv. dissertation, University of California, Los Angeles.

Urish, D.W., and McKenna, T.E. (2004) Tidal effects on ground water discharge through a sandy marine beach. *Groundwater*, 42: 971–982.

Wade T.J., Pai, N., Eisenberg, J.N., and Colford, Jr, J.M. (2003) Do U.S. Environmental Protection Agency water quality guidelines for recreational waters prevent gastrointestinal illness? A systematic review and meta-analysis. *Environmental Health Perspectives*, 111: 1102–1109.

Wait, D.A., and Sobsey, M.D. (2001) Comparative survival of enteric viruses and bacteria in Atlantic Ocean seawater. *Water Science and Technology*, 43: 139–142.

Whitman, R.L., Shively, D.A., Pawlik, H., Nevers, M.B., and Byappanahlli, M.N. (2003) Occurrence of *Escherichia coli* and enterococci in *Cladophora* (Chlorophyta) in nearshore water and beach sand of Lake Michigan. *Applied and Environmental Microbiology*, 69: 4714–4719.

Whitman, R.L., and Nevers, M.B. (2003) Foreshore sand as a source of *Escherichia coli* in nearshore water of a Lake Michigan beach. *Applied and Environmental Microbiology*, 69: 5555–5562.

Whitman, R.L., Nevers, M.B., Korinek, G.C., and Byappanahalli, M.N. (2004) Solar and temporal effects on *E. coli* concentration at a Lake Michigan swimming beach. *Applied and Environmental Microbiology*, 70: 4276–4285.

Wold, S. (1995) PLS for multivariate linear modeling in chemometric methods. In H. van de Waterbeemd (ed.), *Molecular Design Methods and Principles in Medicinal Chemistry*. Verlag-Chemie, Weinheim, Germany.

Wymer, L.J., Brenner, K.P., Martinson, J.W., Stutts, W.R., Schaub, S.A., and Dufour, A.P. (2005) The EMPACT Beaches Project. US Environmental Protection Agency, Office of Research and Development.

11

Statistical sensitivity analysis and water quality

Alessandro Fassò

Department of Information Technology and Mathematical Methods, University of Bergamo, Dalmine BG, Italy

11.1 Introduction

In this chapter, concepts and methods of statistical sensitivity analysis (SA) of computer models are reviewed and discussed in relation to water quality analysis and modeling.

The starting point of this approach is based on modeling the uncertainty of the computer code by probability distributions. Despite the fact that computer models are generally non-stochastic, in the sense that if we rerun the code we get the same result, the stochastic approach turns out to be useful to understand how the input uncertainty is propagated through the computer code into the output uncertainty.

We follow the standard approach to SA, based on variance decomposition, and consider three levels of SA. At the first or preliminary level, we discuss design of experiments (DOE) and response surface methodologies in order to get a first estimate of the input influences on the model output.

At the second level, going further into modeling the relationship between computer model inputs and outputs, we assume that different computer runs are

Statistical Framework for Recreational Water Quality Criteria and Monitoring   Edited by Larry J. Wymer
© 2007 John Wiley & Sons, Ltd

independent. We then discuss techniques derived from Monte Carlo input simulations and regression analysis.

At the third level, recognizing that, since the computer model is actually non-stochastic, the errors are often smoother than independent errors, we consider the geostatistical SA which is based on assuming that the error of the computer code emulator is a stochastic process with positive correlation which gets higher as two inputs get closer.

11.1.1   Computer models and recreational water quality

Computer models are widely used in hydrology and water quality studies in general. In recreational water quality modeling and assessment, the use of both conceptual and management models is becoming increasingly important.

As a first example, consider real-time forecasting of *Escherichia coli* concentrations, which is useful for management beach closure strategies and may be approached by both mechanistic and statistical models. In this framework, Olyphant and Whitman (2004) applied dynamic regression models including hydrological, meteorological and water quality predictors to swimming beaches of Lake Michigan.

Moreover, Vinten *et al.* (2004) compared soil transport models, multiple regression models and distributed catchment models in the catchment of the River Irvine, Scotland. In deep ocean outfall plumes off Sydney, Miller *et al.* (1996) used finite element modeling, to assess both long- and short-term effects.

Reynolds (1999) reviews various modeling strategies for understanding phytoplankton dynamics in water quality and lake management. For river and lake water quality, computer models (CMs) may be used in integrated analyses at the catchment scale where various dimensions are usually taken into account.

In order to assess microbial pollution of rural surface water, Jamieson *et al.* (2004), considered liquid and solid waste generated from industry, zootechny and domestic sources. They review some approaches to modeling both surface and subsurface transport of the associated microorganisms and their flow through stream networks.

Norton *et al.* (2004) considered the hydrologic, economic, and stream sediment sources of uncertainty in a calibrated CM applied to the Ben Chefley Dam, Australia.

Hydrological models are important here because they are often used as submodels of water quality models. For example, Whitehead *et al.* (1997) considered a combined flow- and process-based river quality model including nitrate, dissolved oxygen, biochemical oxygen demand, ammonium ions, temperature, pH and a conservative water quality determinand. In general, mechanistic models have been extensively studied in hydrology, in particular flow models and rainfall – runoff models (see Beven, 2001). In dry areas, such as the Mediterranean, water quality may be severely influenced by reduction in flow. Becciu *et al.* (2000) studied a calibrated conceptual model for minimum instream flow in Central Alps catchments by means of regression modeling and outlier analysis.

Another issue relevant for recreational water is wastewater management. For example, in heavy metal biofilter modeling, Fassò *et al.* (2003) used a conceptual model based on the advection dispersion reaction equation and modeled the multivariate response using a multivariate heteroskedastic statistical approximation.

From the above examples, the CM outputs may be the stream discharge or the concentration of chemicals and/or pollutants or time to next health hazard event; and the CM relates these to anthropic and environmental parameters, initial and boundary conditions, global climate, and dynamics of meteorology.

11.1.2 Sensitivity analysis and chapter organization

Uncertainty may be related to measurement errors, both at model output (MO) and parameter level. Moreover, it may be due to the fact that the CM is only an approximation of the *real system*. Such sources of uncertainty will be discussed in some detail in Section 11.2, where we extend the taxonomy of Kennedy and O'Hagan (2001).

In some cases, the CM needs to be calibrated on some observational data. It is then interesting to assess the estimation or calibration uncertainty and the sensitivity of the MO to the calibration parameters. In other cases, calibration is not explicitly considered, but once again SA is aimed at understanding to what extent the various parameters affect the MO.

Sensitivity analysis is then intended to assess these individual sensitivities and to rank various inputs with respect to certain *sensitivity indexes*. If we avoid uncertainty concepts, the simplest idea for doing SA is to consider first-order local expansion at some *internal point* and use the analytical or numerical partial derivatives to carry out this *local SA*.

In Section 11.3, we discuss the approach known as *global SA*. The aim is to define the global influence of each input to the uncertainty of the MO. Then, using an appropriate global performance measure, such as variance, squared or absolute fitting error or likelihood, we show how to assess and rank the sensitivity to each parameter. We first review and comment on the case considered extensively in Saltelli *et al.* (2000) and in Fassò and Perri (2002) where the CM is taken for granted or, equivalently, no calibration data are available so we assess the sensitivity of the MO without reference to observed data.

In Section 11.4, we discuss the preliminary *SA*, generally based on a reduced number of computer runs and little statistical modeling. In such a case, DOE and response surface methodology techniques are of interest. At a subsequent step, when computer runs are cheap, Monte Carlo SA is useful. This technique and modified sampling strategies (Latin hypercube and importance sampling) are discussed in Section 11.5. In Section 11.6, model-based SA is discussed and the variance-based SA is extended to multivariate and heteroskedastic CMs; in the latter case, the residual model uncertainty is not constant over the input domain.

In Section 11.7, we discuss some SA techniques related to the case where calibration data are available and CM validation may be performed also with SA.

In Section 11.8, we discuss the case where the uncertainty of the MO, prior to running the CM, is assumed to be a stochastic process indexed by the computer model input x. In the previous sections, the Monte Carlo approach was based on independent computer runs. Here, recognizing that the original computer code is non-stochastic, the error smoothness is described by a geostatistical approach.

11.2   Model uncertainty setup

In order to introduce uncertainty concepts, we first suppose that the *true* environmental phenomenon of interest, say ζ, is related to some observable multidimensional inputs $x = (x_1, \ldots, x_k)$ in some input domain, say D, and some other non-observable or unknown inputs x^*, that is,

$$\zeta = \zeta(x, x^*).$$

The computer model or code is a computable function, say $f(x)$, which for given inputs x gives an output

$$z = f(x).$$

Usually it is a complex function and its analytical properties are difficult to derive. In some cases, it may be a stochastic function including, for example, some Monte Carlo or other simulation-based components. In this chapter, we consider mainly deterministic CMs, in the sense that, if we rerun the code, we get the same result. In the simple ideal case the CM is a perfect model so that

$$f(x) = \zeta(x, x^*) \tag{11.1}$$

for every x^*.

11.2.1   Input uncertainty

In environmental CMs it is common to have two kinds of input parameters, that is, fixed and variable parameter vectors denoted by $\theta = (\theta_1, \ldots, \theta_h)$ and $x = (x_1, \ldots, x_k)$ respectively, giving the CM equation

$$z = f(x, \theta). \tag{11.2}$$

The vector θ is often referred to as the 'calibration parameter' to be *estimated* on observational data. For example, in a hydrological model applied to a certain watershed, the parameter set θ may be related to geomorphological and/or evapotranspiration parameters of that watershed, while $x = (t, y_1, y_2)$ may be the time index $t = 1, 2, \ldots$ and meteorological conditions y_1 and discharges y_2 at time t.

We are often interested in the global behaviour of the true system ζ without fixing the input x. Or in a risk analysis, we are interested in right-tail behavior of risk-related MOs. So, in practice, the *k-dimensional* input $x = (x_1, \ldots, x_k)$ is uncertain and it may be useful to describe such uncertainty by an appropriate *k-variate* probability distribution with joint probability density function given by $p(x)$ and cumulative distribution $P(x)$.

The simplest example of input distribution is given by independent rectangular marginals. We will see in the following sections that when inputs are independent the sensitivity indexes satisfy certain additivity properties.

In some cases this simple setup has to be replaced by other multivariate distribution. For example, in Fassò (2006, Section 2.1), considering the SA of a heavy metal biofilter CM, the maximum uptake constant (q_{max}) and the Langmuir constant (b) are supposed bivariate normal with moderate positive correlation, $\rho = 0.30$, to reflect the calibration uncertainty source of these parameters.

11.2.2 Simulation and residual uncertainty

Except in the simplistic case of equation (11.1), since x^* is unknown, the CM or simulator $z = f$ is, at best, an approximation of the averaged values of ζ, say μ. This is given by

$$\mu(x) = E_{x^*}(\zeta(x, x^*)|x),$$

where $E_{x^*}(.|x)$ is the conditional expectation operator with respect to some conditional distribution $p(x^*|x)$. Hence, if ζ is observed without error, the *residual uncertainty* is given by the probability distribution of

$$e_0 = \zeta - \mu \tag{11.3}$$

and the CM inadequacy or simulator uncertainty is given by

$$e_1 = f - \zeta = \bar{e}_1 + e_0, \tag{11.4}$$

where $\bar{e}_1 = f - \mu$ is the *partial simulation uncertainty* and e_1 is the *total simulation uncertainty*. If observational data, say Z, are available about ζ then measurement errors are possible and

$$Z = \zeta + \varepsilon_\zeta.$$

This case may be handled in the Bayesian framework of Section 11.8 or, under Markovian assumptions on the unobserved ζ, by the dynamical system setup and the Kalman filter; see Fassò and Nicolis (2005) for an application of this approach to air quality.

11.2.3   Emulation

The next step is to suppose that we have a simplified model, say $g(x, \beta)$, where β is a *regression-type* parameter to be estimated in order to give a *good* approximation of the CM, $f(x)$. Of course, we have partial and total emulation uncertainty given, respectively, by

$$\bar{e}_2 = g - f \tag{11.5}$$

and

$$e_2 = g - \zeta = \bar{e}_2 + e_1.$$

The fixed but unknown parameter β may be interpreted, for example, as the minimum mean square error parameter which minimizes

$$E_x \left(g(x, \beta) - f(x) \right)^2.$$

11.2.4   Estimated emulator

In practice we may get an estimate $\hat{\beta}$ using simulated data from the CM

$$(x_i, f(x_i)), \qquad i = 1, \ldots, n.$$

This gives the estimated emulator

$$\hat{g}(x) = g \left(x, \hat{\beta} \right),$$

and we have another two sources of uncertainty, say partial and total estimation uncertainty, given respectively by

$$\bar{e}_3 = \hat{g} - g$$

and

$$e_3 = \hat{g} - f. \tag{11.6}$$

In some cases $\hat{\beta}$ is a statistical estimate, for example a maximum likelihood estimate, and the uncertainty in β and the errors e_3 and $\bar{e}_3$ may be assessed using some standard approximate normality and confidence intervals. In other cases $\hat{\beta}$ is calibrated using, for example, hydrological techniques giving involving generatized likelihood uncertainty estimation, which is discussed in Section 11.7.1.

In Sections 11.4 and 11.5, the quantities of main interest are the emulated values $\hat{g}$ and the corresponding errors given by equation (11.6).

11.2.5 Output uncertainty

The uncertainty in the input x propagates to the output z via the CM, $f(x)$, so that, as long as x is a random vector with distribution $p(x)$, we are interested in the output uncertainty distribution, $p(z)$ say, which is related to $p(x)$ via the code $f(x)$. For example, in risk analysis we are interested in the cumulative output distribution $P(z)$ and its right-tail quantiles.

A typical quantity for assessing the squared uncertainty is the output variance, which may be computed using the input uncertainty distribution $p(x)$:

$$\mathrm{Var}(z) = \sigma_z^2 = \int_D (f(x) - f_0)^2 p(x) dx,$$

where $f_0 = E(z)$.

Moreover, the MOs may be compared with observational data of the *true system*. Let e be the *forecasting error* according to one of the setups from Sections 11.2.1–11.2.4. For example, for an exactly observed system with known CM, we have $e = f - \mu$ and, for an emulated model, the forecasting error is $e = \hat{g} - \mu$.

As can be seen from these last two quantities, such error accounts also for model inadequacy. Therefore, the output uncertainty is generally given by the error cumulative distribution, $P(e)$, say. If we have replicated input values, for example a random sample $x_1, \ldots, x_n$ from $p(x)$ as discussed in Section 11.5, we can use standard statistical inference to estimate $P(e)$, its mean, variance, confidence intervals, etc.

11.3 Variance-based sensitivity analysis

Most of the remaining part of this chapter is based on data coming exclusively from the CM. Hence, except in section 11.7, we will not consider in detail either the residual uncertainty (11.3) or the simulator uncertainty (11.4).

In principle the sensitivity of the MOs, z, to each component of $x = (x_1, \ldots, x_k)$ may be based on the local approach by the partial derivatives, $\partial f / \partial x_j$, which can be computed either analytically or numerically around a 'central point' $x^0 = (x_1^0, \ldots, x_k^0)$. Although this approach has been used for a long time and is still being used, it is rather simplistic for complex nonlinear CMs.

Extending the local SA to 'many' $x^0 \in D$ would give more information but, of course, would rebuild the complexity and the multidimensionality of f itself. So we need a 'global' approach that is able to give information for every x but is also a synthesis which reduces the original complexity. Moreover, we seek quantities that can be 'estimated' on a reduced set of CM runs.

The basic idea of global SA is to study the overall influence of each input component x_j on the uncertainty of the MO. In variance-based SA, we assess the uncertainty by the variance and we are naturally led to SA measures based on variance decomposition, for example using a main-effect model

$$z = f_0 + \sum_{j=1}^{k} f_j + \varepsilon, \tag{11.7}$$

with $f_0 = E(z)$ as above and

$$f_j = E(z|x_j) - f_0.$$

Note that the error ε here is non-stochastic as it is a pure model-inadequacy quantity. Whenever the standard statistical interpretation does not hold, in many situations such an error, being a complicated function of many independent inputs x, behaves close to a stochastic error.

If the inputs are independent we can decompose the total uncertainty as

$$\mathrm{Var}\,(z) = \sum_{j=1}^{k} \mathrm{Var}\,(f_j) + \mathrm{Var}\,(\varepsilon), \tag{11.8}$$

and Pearson's correlation ratio

$$S_j = \frac{\mathrm{Var}\,(f_j)}{\sigma_z^2} \tag{11.9}$$

is the *natural* first-order sensitivity index for x_j. As a matter of fact, $\mathrm{Var}(f_j)$ may be interpreted as that part of the uncertainty of the output which can be reduced by fixing the jth input parameter and, correspondingly, the sensitivity S_j may be interpreted as the fraction of (squared) uncertainty of z due to the uncertainty on X_j.

In principle, we can assess interactions of any order starting from the full interaction model

$$z = f_0 + \sum_j f_j + \sum_{i<j} f_{i,j} + \ldots + f_{j_1,\ldots,j_k}, \tag{11.10}$$

where $f_{i,j} = E\,(z|x_i, x_j) - f_i - f_j - f_0$ and so on. In this case, in order to cover the effect of the interactions between x_j and the other inputs, the sensitivity index S_j may be increased to get the total effect. To see this, let $x_{(j)}$ be the $(k-1)$-dimensional vector corresponding to x without the jth component, and consider the decomposition

$$z = f_0 + f_j + f_{(j)} + f_{j,(j)}.$$

Now, using the input independence, we have

$$\mathrm{Var}(z) = \mathrm{Var}(f_j) + \mathrm{Var}(f_{(j)}) + \mathrm{Var}(f_{j,(j)})$$

and, following Homma and Saltelli (1996), the total sensitivity index for x_j is given by

$$S_{T_j} = \frac{\mathrm{Var}(f_j) + \mathrm{Var}(f_{j,(j)})}{\sigma_z^2} = 1 - S_{(j)}.$$

11.3.1 Further details

Let D_j and $D_{(j)}$ be the input domains of x_j and $x_{(j)}$ respectively. Then the output response to x_j is given by the $(k-1)$-dimensional integral

$$E(z|x_j) = \int_{D_{(j)}} f(x)p(x_{(j)})dx_{(j)} \qquad (11.11)$$

and its variance is given by the one-dimensional integral

$$\mathrm{Var}(E(z|x_j)) = \int_{D_j} E(z|x_j)^2 p(x_j)dx_j - f_0^2. \qquad (11.12)$$

Moreover, $\mathrm{Var}\left(f_{(j)}\right)$, which enters the total sensitivity S_{T_j}, is given by

$$\mathrm{Var}(E(z|x_{(j)})) = \int_{D_{(j)}} E(z|x_{(j)})^2 p(x_{(j)})dx_j - f_0^2.$$

Shortcuts for estimating $S_{(j)}$ are discussed in Chan *et al.* (2000), and their efficiency may be assessed in practice using equation (27) of Fassò and Perri (2002).

11.4 Exploratory analysis

At the early stages of the CM analysis, especially if the computer runs are expensive, it may be worth considering a simplified emulator, based on a reduced set of values for each input x_j. For example, consider just binary inputs which assume the values high/low, giving the new input domain, say D^*, with 2^k different values. This is known as a 2^k factorial design; it requires the code to be run 2^k times and allows one to identify the zero-error full interaction model (11.10).

When the figure 2^k is too large and/or there are too many interactions, we need *smaller* designs such as the fractional 2^{k-h} designs which allow the estimation of a reduced version of model (11.10) with high-order interactions being encompassed in the error component as in model (11.7).

The related techniques known as design of experiments and response surface methodology are well established for stochastic experiments; see the classic Box *et al.* (1978) and the more recent Wu and Hamada (2000). Sacks *et al.* (1978)

considered the so-called *design of computer experiments*, and its optimization is also discussed in Section 11.8.1. Although, at first sight, standard DOE seems to work also in this case, it has to be recognized that, due to the non-stochastic nature of computer experiments, replications, blocking, and randomization now lose their usual meaning.

Moreover, for output uncertainty estimation, considering only binary inputs may be of limited value. One can extend to the n-level factorial design with n^k components or fractionally reduced, but as n and k are not very small it does not work in practice.

11.5   Monte Carlo and other sampling techniques

In the previous section 'optimal' systematic sampling was considered for the case where the response surface is fixed in advance and certain cardinality reduction assumptions are in order. Using this approach for complex emulators $g(x)$ and/or high-dimensional and high-cardinality inputs is not feasible in practice because of computational complexity. In this section, therefore, we discuss some techniques which are not optimal but which are informative for any particular emulator.

To do this, we use the idea of Section 11.2.1 which describes the input uncertainty by a certain probability distribution, say $P(x)$, with the assumption that different runs are independent. This gives a natural way to obtain information about the CM, that is, simple random sampling from $P(x)$. This means that we need (pseudo-)random numbers from $P(x)$, which is easily taken care of with standard software. Using this approach we get a (possibly large) sample from the CM, namely $(x_1, z_1), \dots, (x_n, z_n)$ which is informative about the code f and may be used for empirical modeling, estimation, and validation of the emulator g. Moreover, it is useful for estimating the indexes of Section 11.3 and, as $z_1, \dots, z_n$ are a random sample from the unknown distribution $P(z)$, it may be used to get the estimated output uncertainty distribution, say $\hat{P}(z)$.

Of course this approach is especially appropriate when computer runs are cheap and getting 'a large Monte Carlo sample' is a feasible task in terms of computing resources.

11.5.1   Importance sampling

Suppose we are interested in estimating the average of the positive output function $h(x) > 0$:

$$\mu = E(h(x)) = \int_D h(x)p(x)dx.$$

For example, we may be interested in computing the output mean, with $h(x) = |f(x)|$ or the variance with $h(x) = (f(x) - f_0))^2$. Using the standard Monte Carlo approach, we would estimate μ by means of a random sample $x_1, \ldots, x_n$ from $p(x)$ and its sample average

$$m = \frac{1}{n} \sum h(x_i).$$

The idea of importance sampling is to use a stratified sample from a cumulative distribution $Q(x) \neq P(x)$ which gives higher probability to those inputs x, where $h(x)$ is large. In practice the ith stratified importance sample, x_i', say, is given by

$$x_i' = Q^{-1} \left(\frac{i - 1 + R_i}{n} \right), \tag{11.13}$$

where R_i is a uniform random number and the unknown μ is now estimated by the unbiased weighted estimator

$$m' = \sum \frac{h(x_i')}{q(x_i')/p(x_i')}. \tag{11.14}$$

It is easily seen that if $q(x) = h(x)p(x)/\mu$ then m' has zero variance and, hence, is optimal. On the one hand, the sampling strategy, which increases the sampling size where the CM uncertainty is large, is more efficient than standard Monte Carlo sampling. On the other hand, application of this method requires approximate knowledge of the CM itself. Moreover, in equation (11.14), weighting is essential to avoid bias. Finally, if x is multivariate then stratified sampling gives the curse of dimensionality of the previous section and Latin hypercube sampling (discussed next) should be taken into consideration.

11.5.2   Latin hypercube sampling

Latin hypercube sampling (LHS) is a multidimensional generalization of stratified sampling which assigns each scalar sample x_i, $i = 1, \ldots, n$, to a different equiprobability interval or cell c_i, say, using equation (11.13) with P_i instead of Q. In the k-dimensional case, we have a k-dimensional grid of n^k cells c_i given by the Cartesian product of the marginal intervals $c_{i,j}$, that is $c_i = c_{i,1} \times \ldots \times c_{i,k}$.

The n^k factorial design of previous sections would simply give one element x_i for each cell c_i. Now in *LHS*, as shown by Figure 11.1, the cells are chosen so that each marginal has just one observation in each of the n equiprobability intervals and it may be seen as a highly fractionalized factorial design. As a matter of fact, the term comes from Latin squares where there is an array of symbols and each occurs just once.

Algorithm

To do this, note that the cells c_i are identified by k integers ranging from 1 to n. Hence the $n \times k$ matrix C of such integers has columns which are given by random

	X			
		X		
		X		
X				
			X	

Figure 11.1   Example of a two-dimensional Latin hypercube assignment with $n=5$ and rectangular marginals.

permutations of the integers $1, \ldots, n$. After choosing the cell c_i the value x_i is chosen from $P(x|c_i)$ using equation (11.13), thanks to independence. The extension to certain correlation structures is considered by Stein (1987).

Optimality

It is known that if the code $f(x)$ is monotonic in each component x_j, then LHS improves on random sampling for estimating the output mean, variance, and cumulative distribution function (McKay *et al.*, 1979). Nevertheless, due to the high degree of fractionalization, this technique requires some caution when used for high-order interaction models (e.g. Huntington and Lyrintzis, 1998).

11.6   Model-based sensitivity analysis

In Section 11.4, we considered response surface methodology as a way to understand how the inputs affect the computer code. In this section, we are more deeply concerned with the model emulator and its capability to give further insight into the CM in general and in its sensitivity indexes in particular.

Let us start by considering a linear regression emulator

$$g(x, \beta) = \beta_0 + \sum_{j=1}^{k} x_j \beta_j \qquad (11.15)$$

having errors (11.5) that are approximately independent, homoskedastic Gaussian. If the input components are uncorrelated as in Section 11.3, we get the sensitivity indexes S_j from the variance decomposition,

$$\sigma_z^2 = \sum_{j=1}^{k} \sigma_{x_j}^2 \beta_j^2 + \sigma_\varepsilon^2. \qquad (11.16)$$

To do this, using a large enough Monte Carlo sample, we can use the least-square estimates of β to get the estimated sensitivity indexes

$$\hat{S}_j = \frac{\hat{\sigma}_{x_j}^2 \hat{\beta}_j^2}{\hat{\sigma}_y^2},$$

and from (11.16) we have

$$\sum \hat{S}_j = R^2 = 1 - \frac{\hat{\sigma}_\varepsilon^2}{\hat{\sigma}_y^2}. \qquad (11.17)$$

This approach easily extends to interactions, polynomial components, and transformed inputs, using, for example the generalized linear model

$$g(x, \beta) = h(x)'\beta. \qquad (11.18)$$

Some caution is required for high-dimensional input sets and high-order interactions. For example, Helton *et al.* (2005), carrying out SA of a waste isolation plant with more then 30 inputs, found that stepwise regression was unstable and they preferred separated analyses.

11.6.1 Nonlinear and multivariate sensitivity analysis

Often the code output is a vector and we are interested in assessing the sensitivity of the CM as a whole. For example, considering a wastewater biofilter model, Fassò *et al.* (2003) were interested in performance outputs given by the length of unused biofilter bed as well as the breakthrough time, which is the working time over which it is necessary to regenerate the fixed bed. In this case, using the covariance decomposition which extends equation (11.16) to the multivariate case, they proposed both the trace sensitivity indexes which retain additivity as in equation (11.17) and determinantal sensitivity indexes which consider also the output correlations.

Nonlinear extensions of the linear model (11.15) follow two main approaches. Keeping homoskedastic independent errors, the first path focuses on generalizing the parametric emulator into nonparametric models. In the case of additive models and independent inputs, the decomposition (11.8) and the sensitivity indexes (11.9) may be still used.

The second nonlinearity approach arises when the emulator errors (11.6) are heteroskedastic and the output uncertainty depends on certain input parameters. For example, continuing with the above biofilter example, it has been found that the emulator errors for the length of unused biofilter bed may be modeled as

$$e_3 = \varepsilon\sqrt{\alpha_0 + \alpha_1 u + \alpha_2 u^2}, \qquad (11.19)$$

where ε is a standardized error with unit variance and u is the input parameter given by adsorption particle diameter. Equation (11.19) shows that the model uncertainty is not constant over the input domain D and the model predictions are more reliable for certain input values.

The sensitivity indexes may account easily for heteroskedasticity. In the biofilter case, extending equation (11.16) for heteroskedasticity, the index for the adsorption particle diameter is given by

$$\hat{S}_u = \frac{\hat{\sigma}_u^2 \hat{\beta}_u^2}{\hat{\sigma}_y^2} + \frac{\alpha_1 \hat{E}(u) + \alpha_2 \hat{E}(u^2)}{\hat{\sigma}_y^2}.$$

Note that the second term on the right-hand side is part of the residual uncertainty $\text{Var}(e_3)$ which, thanks to the heteroskedastic approach, has been attributed to the adsorption particle diameter.

11.7 Sensitivity analysis and calibration

Often a CM, being in the form of equation (11.2), requires appropriate calibration and validation on some observational data sets. For example, Sincock *et al.* (2003) considered a river water quality model under unsteady flow conditions including a flow component and a water quality component. After calibration on historical data they found that the model performance was insensitive to algal activity while nitrification and sedimentation were important.

We will not go much further into validation issues here; we only remark that one of the steps in validation is the understanding of the performance of the CM with respect to variation of fixed parameters. For example, if the model performance is not sensitive to a parameter component θ_j then the observational data are inappropriate for that parameter or the CM is overparameterized for that application.

11.7.1 Equifinality and generalized likelihood uncertainty estimation

Hydrological modeling often requires some form of calibration so that the fixed CM parameters θ in (11.2) are adjusted to get a better fit to some observed data. In this section, we consider methods developed in hydrology, but useful beyond that for various instances of CM calibration and validation. For example, McIntyre and Wheater (2004) considered the calibration of a simulation model for monthly total phosphorus in the Hun River, China.

Using the notation and concepts of section 11.2, we then have a set of observed data $(x_1, \zeta_1), \dots, (x_N, \zeta_N)$ and we want to understand the influence of the parameter vector $\theta = (\theta_1, \dots, \theta_n)$ on the forecasting performance of the CM with respect to this data. Such performance is traditionally based on the mean of squared errors

$$\hat{\sigma}^2_{e(\theta)} = \frac{1}{N} \sum_{i=1}^{N} (\zeta_i - f(x_i, \theta))^2,$$

but other measures may be used, such as the mean of absolute errors (MAE) or maximum relative error.

We then have the so-called *likelihood measure*, L say, discussed by Beven (2001), which is constrained to be zero for *non-behavioral* values of θ and one for the ideal case of perfect forecasts $\zeta_i = f(x_i, \theta)$. The first example is the truncated forecasting efficiency

$$L(\theta) = \begin{cases} 1 - \dfrac{\hat{\sigma}^2_{e(\theta)}}{\hat{\sigma}^2_{\zeta}}, & \textit{if } L > 0, \\ 0 & \text{otherwise}, \end{cases}$$

which is well known to statisticians as the coefficient of determination R^2. A second example is the Box and Tiao measure

$$L(\theta) = (\hat{\sigma}^2_{e(\theta)})^{-H},$$

where $H > 0$ is a subjective shaping coefficient.

Equifinality arises here since it is common in environmental applications that $L(\theta)$ is almost the same for many different values of θ. In other words, we have the well-known modeling fact that different CMs give forecasts which are almost the same with respect to a certain likelihood measure L.

Hence, a natural choice is to apply output uncertainty to the new CM given by $f(x, \theta)$ weighted by the likelihood measure. To do this, consider n Monte Carlo simulations, $\theta_1^*, \ldots, \theta_n^*$ of the possibly multivariate parameter $\theta = (\theta_1, \ldots, \theta_h)$ with rectangular marginal distributions and consider the normalized likelihood

$$\bar{L}(\theta_{i*}) = \frac{L(\theta_i^*)}{\sum_{i=1}^{n} L(\theta_i^*)}.$$

Now, suppose that the quantity of interest is a function $Q = Q[f(x, \theta)]$ with weighted Monte Carlo cumulative distribution given by

$$\hat{F}(q) = \hat{P}(Q \leq q) = \sum_{i:Q(\theta_i^*) \leq q} \bar{L}(q_i^*) \tag{11.20}$$

For example, if $Q = z$, equation (11.20) allows the computation of the weighted forecasting quantiles. Moreover, if Q is the ith component of θ, namely $Q = \theta_i$, (11.20) gives the marginal cumulative distribution of θ_i. Hence, SA may be performed on

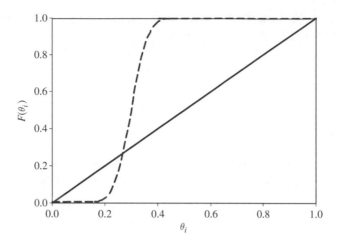

Figure 11.2   Weighted cumulative distribution of $Q = \theta_i$, $\theta \sim$ Rectangular(0,1), $\hat{F}(\theta_i) = \sum_{j:\theta_{ij}^* \leq \theta_i} \bar{L}(\theta_i)$.

graphical grounds by comparing this marginal with the uniform distribution which may be interpreted as the prior Monte Carlo distribution. In particular, for a hypothetical example, Figure 11.2 shows the reduction in output uncertainty achievable in θ_i by multivariate calibration of θ.

11.8   Geostatistical sensitivity analysis

So far, we have used methods that assume independence of emulator errors between computer runs. In this section, we consider methods which imply more complex modeling and computing time. Hence, they are appropriate for cases where the CM is an 'expensive function' and large Monte Carlo computer experiments are not feasible. Moreover, this approach is efficient when we are dealing with a 'smooth CM', where smoothness means that $f(x)$ and $f(x')$ are highly correlated for x close to x'.

The basic idea of Oakley and O'Hagan (2004) is to consider the model output $f(\mathrm{x})$ as a stochastic process indexed by the CM input x in the sense that, for a fixed hypothetical sequence of inputs, say $x_1, \ldots, x_n$, the model outputs, namely $f(x_1), \ldots, f(x_n)$, are correlated random variables. This stochastic process representation may be interpreted as Bayesian beliefs about the MOs prior to running the code.

Whenever x is assumed to be non-stochastic, it is considered to be unknown with uncertainty distribution $p(x)$. This approach with $x \in D$ can be seen as a geostatistical approach and, in this sense, we will use terms such as *space* for D. It follows that the sensitivity quantities introduced in Section 11.3 are stochastic quantities, for example the *spatial* averages (11.11) and variances (11.12) are integrals of a stochastic process. Given a set of MOs $(x_1, z_1), \ldots, (x_n, z_n)$, the above spatial integrals can be estimated by the posterior counterparts of $f(x)$. For example, suppose that $\hat{f}(x)$ is an appropriate Bayesian kriging estimate of $f(x)$ given by

$$\hat{f}(x) = E\left(f(x)|z_1, \ldots z_n\right).$$

Then the spatial average (11.11) is estimated by

$$E^*(z|x_j) = \int_{D_{(j)}} \hat{f}(x)p(x_{(j)})dx_{(j)}$$

and similarly, the spatial variance (11.12) by

$$\mathrm{Var}^*(z|x_j) = \int_{D_{(j)}} \hat{f}(x)^2 px_{(j)}dx_{(j)} - \hat{f}_0^2.$$

To do this, the prior uncertainty on the model output $f(x)$ before actually running the CM is modeled by a Gaussian stochastic process with mean value given by

$$E(f(x)|\beta) = h(x)'\beta,$$

where $h(x)$ is a known input transformation as in equation (11.18) and β is a hyperparameter. The covariance function of $f(x)$ is given by

$$\mathrm{Cov}(f(x), f(x')|\sigma^2) = \sigma^2 c(x, x'),$$

where $c(x, x')$ is a geostatistical correlation function; for example, in the stationary isotropic case, we have

$$c(x, x') = c(|x - x'|).$$

Moreover, $c(0) = 1$ and $c(t)$ decreases with increasing t and, in general, may depend on some further hyperparameters, say γ.

If γ is known and the hyperparameters (β, σ^2) have prior

$$p\left(\beta, \sigma^2\right) \alpha \sigma^2, \tag{11.21}$$

then $\hat{f}$ and $\hat{c}$ have closed-form representation and, marginally to (β, σ^2), the MOs have a multivariate t distribution. In particular,

$$t(x) = \frac{f(x) - \hat{f}(x)}{\sqrt{\hat{c}(x, x')}} \tag{11.22}$$

has a t distribution with $k + n$ degrees of freedom.

If the prior distribution is not as in (11.21) or γ is unknown, the closed-form posterior distribution (11.22) does not hold, and Markov chain Monte Carlo integration is required, resulting in a considerably greater computational burden. To avoid this, it is common practice in Bayesian kriging to use a plug-in approach based on substituting the posterior estimate for γ, say $\hat{\gamma}$, into $c(x, x')$ and, conditionally on this, use the above methods.

11.8.1   DOE

In this framework, the input design is different from the Monte Carlo approach of Section 11.5 because, here, x is non-stochastic but the integrals to be estimated are stochastic ones. As a matter of fact, Sacks *et al.* (1989) discuss the extension of the classical DOE of Section 11.4 to DOE for stochastic processes. In general terms, it is based on the optimization of the integrated mean squared error

$$IMSE(x_1, \ldots, x_n) = \int_D E\left[\left(f(x) - \hat{f}(x) \right)^2 | x_1, \ldots, x_n \right] p(x) dx,$$

giving both sequential and non-sequential design algorithms. Since the MOs are not independent, algorithms are non-standard and may be time-consuming. Of course this is worthwhile if the computer runs are more expensive.

Acknowledgment

The work for this chapter was partially supported by Italian MIUR-Cofin 2004 grants.

References

Becciu, G., Bianchi, A., Fassò, A., Fassò, C.A., and Larcan, E. (2000) Quick calculation of minimum in-stream flow in drainage basins of Central Alps. In U. Maione, B. Majone-Lehto, and R. Monti (eds) *New Trends in Water and Environmental Engineering for Safety and Life*. Balkema, Rotterdam.

Beven, K.J. (2001) *Rainfall-runoff modelling*. John Wiley & Sons, Ltd, Chichester.

Box, G.E.P., Hunter, W., and Hunter J. (1978) *Statistics for Experimenters. An Introduction to Design, Data Analysis and Model Building*. John Wiley & Sons, Inc., New York.

Chan, K., Tarantola, S., Saltelli, A., and Sobol', I. (2000) Variance-based methods. In A. Saltelli, K. Chan, and M. Scott (eds) *Sensitivity Analysis*. John Wiley & Sons, Ltd, Chichester.

Fassò, A. (2006) Sensitivity analysis for environmental models and monitoring networks. In A. Voinov, A.J. Jakeman, and A.E. Rizzoli (eds), *Proceedings of the iEMSs Third Biennial Meeting, 'Summit on Environmental Modelling and Software'*. International Environmental Modelling and Software Society, Burlington, USA, July. http://www.iemss.org/iemss2006/sessions/all.html.

Fassò, A., and Nicolis, O. (2005) Space-time integration of heterogeneous networks in air quality monitoring. In *Proceedings of the Italian Statistical Society Conference on 'Statistica e Ambiente'*, Messina, 21–23 September, Vol. 1.

Fassò, A., and Perri, P.F. (2002) Sensitivity Analysis. In A. El-Sharaawi and W. Piegorsch (eds), *Encyclopedia of Environmetrics*, Vol. 4, pp. 1968–1982. John Wiley & Sons, Ltd, Chichester.

Fassò, A., Esposito, E., Porcu, E., Reverberi, A.P., and Vegliò, F. (2003) Statistical sensitivity analysis of packed column reactors for contaminated wastewater. *Environmetrics*, 14: 743–759.

Helton, J.C., Davis, F.J., and Johnson, J.D. (2005) A comparison of uncertainty and sensitivity analysis results obtained with random and Latin hypercube sampling. *Reliability Engineering and System Safety*, 89: 305–330.

Homma, T., and Saltelli, A. (1996) Importance measures in global sensitivity analysis of nonlinear models. *Reliability Engineering and System Safety*, 52, 1–17.

Huntington, D.E., and Lyrintzis, C.S. (1998) Improvements to and limitations of Latin hypercube sampling. *Probabilistic Engineering Mechanics*, 13: 245–253.

Jamieson, R., Gordon, R., Joy, D., and Lee, H. (2004) Assessing microbial pollution of rural surface waters: A review of current watershed scale modeling approaches, *Agricultural Water Management*, 70(1): 1–17.

Kennedy, M.C., and O'Hagan, A. (2001) Bayesian calibration of computer models, *Journal of the Royal Statistical Society B*, 63: 425–464.

McKay, M.D., Beckman, R.J., and Conover, W.J. (1979) A comparison of three methods for selecting values of input variables in the analysis of output from a computer code. *Technometrics*, 21: 239–245.

McIntyre, N.R., and Wheater, H.S. (2004) Calibration of an in-river phosphorus model: prior evaluation of data needs and model uncertainty. *Journal of Hydrology*, 290: 100–116.

Miller, B.M., Peirson, W.L., Wang, Y.C., and Cox, R.J. (1996) An overview of numerical modelling of the Sydney deepwater outfall plumes. *Marine Pollution Bulletin*, 33: 147–159.

Norton, J.P., Newham, L.T., and Andrews, F.T. (2004) Sensitivity analysis of a network-based, catchment-scale water-quality model. In *Transactions of the 2nd Biennial Meeting of the International Environmental Modelling and Software Society*, iEMSs: Manno, Switzerland.

Oakley, J., and O'Hagan, A. (2004) Probabilistic sensitivity analysis of coplex models: a Bayesian approach. *Journal of the Royal Statistical Society B*, 66: 751–769.

Olyphant, G.A., and Whitman, R. (2004) Elements of a predictive model for determining beach closures on a real time basis: the case of 63rd Street Beach Chicago. *Environmental Monitoring and Assessment*, 98: 175–190.

Reynolds, C.S. (1999) Modelling phytoplankton dynamics and its application to lake management, *Hydrobiologia*, 395: 123–131.

Sacks, J., Welch, W.J., Mitchell, T.J., and Wynn, H.P. (1989) Design and analysis of computer experiments, *Statistical Science*, 4: 409–423.

Saltelli, A., Chan, K., and Scott, M. (2000) *Sensitivity Analysis*. John Wiley & Sons, Inc., New York.

Sincock, A.M., Wheater, S.S., and Whitehead, P.G. (2003) Calibration and sensitivity analysis of a river water quality model under unsteady flow conditions. *Journal of Hydrology*, 277: 214–229.

Stein, M.L. (1987). Large sample properties of simulations using Latin hypercube sampling. *Technometrics*, 29(2): 143–151.

Vinten, A.J.A., Lewis, D.R., McGechan, M., Duncan, A., Aitken, M., Hill, C., and Crawford, C. (2004) Predicting the effect of livestock inputs of *E. coli* on microbiological compliance of bathing waters. *Water Research*, 38: 3215–3224.

Whitehead, P.G., Williams, R.J., and Lewis, D.R. (1997) Quality simulation along river systems (QUASAR): model theory and development. *Science of the Total Environment*, 194/195, 447–456.

Wu, J., and Hamada, M. (2000) *Experiments: Planning, Analysis and Parameter Design Optimization*. John Wiley & Sons, Inc., New York.

Index

STATISTICS IN PRACTICE

Human and Biological Sciences

Berger – Selection Bias and Covariate Imbalances in Randomized Clinical Trials
Brown and Prescott - Applied Mixed Models in Medicine, Second Edition
Chevret (Ed) – Statistical Methods for Dose Finding Experiments
Ellenberg, Fleming and DeMets – Data Monitoring Committees in Clinical Trials: A Practical
    Perspective
Hauschke, Steinijans and Pigeot – Bioequivalence Studies in Drug Development: Methods
    and Applications
Lawson, Browne and Vidal Rodeiro – Disease Mapping with WinBUGS and MLwiN
Lui – Statistical Estimation of Epidemiological Risk
*Marubini and Valsecchi - Analysing Survival Data from Clinical Trials and Observation
    Studies
Molenberghs and Kenward – Missing Data in Clinical Studies
O'Hagan – Uncertain Judgements: Eliciting Experts' Probabilities
Parmigiani – Modeling in Medical Decision Making: A Bayesian Approach
Pintilie – Competing Risks: A Practical Perspective
Senn – Cross-over Trials in Clinical Research, Second Edition
Senn – Statistical Issues in Drug Development, Second Edition
Spiegelhalter, Abrams and Myles – Bayesian Approaches to Clinical Trials and Health-Care
    Evaluation
Whitehead - Design and Analysis of Sequential Clinical Trials, Revised Second Edition
Whitehead – Meta-Analysis of Controlled Clinical Trials
Willan – Statistical Analysis of Cost-effectiveness Data
Winkel and Zhang – Statistical Development of Quality in Medicine

Earth and Environmental Sciences

Buck, Cavanagh and Litton – Bayesian Approach to Interpreting Archaeological Data
Glasbey and Horgan – Image Analysis in the Biological Sciences
Helsel – Nondetects and Data Analysis: Statistics for Censored Environmental Data
McBride – Using Statistical Methods for Water Quality Management
Webster and Oliver – Geostatistics for Environmental Scientists, 2nd Edition
Wymer – Statistical Framework for Recreational Water Quality Criteria and Monitoring

Industry, Commerce and Finance

Aitken - Statistics and the Evaluation of Evidence for Forensic Scientists, Second Edition
Balding - Weight-of-evidence for Forensic DNA Profiles
Lehtonen and Pahkinen - Practical Methods for Design and Analysis of Complex Surveys,
    Second Edition
Ohser and Mücklich - Statistical Analysis of Microstructures in Materials Science
Taroni, Aitken, Garbolino and Biedermann - Bayesian Networks and Probabilistic Inference
    in Forensic Science

*Now available in paperback